XXV-th Congress on Spectroscopy 2016

EPJ Web of Conferences Volume 132 (2017)

Moscow, Russia
3 - 7 October 2016

Editors:

Evgeny A. Vinogradov
Andrei V. Naumov
Maxim G. Gladush
Kamil R. Karimullin

ISBN: 978-1-5108-3369-2

Printed from e-media with permission by:

Curran Associates, Inc.
57 Morehouse Lane
Red Hook, NY 12571

Some format issues inherent in the e-media version may also appear in this print version.

Printed by Curran Associates, Inc. (2016)

For additional information, please contact EDP Sciences – Web of Conferences
at the address below.

EDP Sciences – Web of Conferences
17, Avenue du Hoggar
Parc d'Activité de Courtabœuf
BP 112
F-91944 Les Ulis Cedex A
France

Phone: +33 (0) 1 69 18 75 75
Fax: +33 (0) 1 69 28 84 91

contact-edps@webofconferences.org

Additional copies of this publication are available from:

Curran Associates, Inc.
57 Morehouse Lane
Red Hook, NY 12571 USA
Phone: 845-758-0400
Fax: 845-758-2633
Email: curran@proceedings.com
Web: www.proceedings.com

TABLE OF CONTENTS

XXV-th Congress on Spectroscopy: preface

A.V. Naumov[1-3,*], K.R. Karimullin[1,2], and E.A. Vinogradov[1,3]

[1]Institute for Spectroscopy RAS, 108840 Moscow, Troitsk, Russia
[2]Moscow State Pedagogical University, 119991 Moscow, Russia
[3]Moscow Institute of Physics and Technology, 142700 Dolgoprudny, Russia

The Congress on Spectroscopy is a regular quadrennial scientific meeting. The series was started in the Soviet Union in 1930-th and has been carefully continued. Nowadays the Congress brings together participants from Russia, neighboring countries (the former Soviet republics) and guests from other countries worldwide. Traditionally, the Congress aims at giving an overview of all areas of modern optics and spectroscopy as well as the latest and topical achievements in these and related disciplines. In previous years, the main organizer of this meeting was the Institute for Spectroscopy of the Russian Academy of Sciences (www.isan.troitsk.ru).

The XXV-th Congress on spectroscopy was held in Troitsk city district of Moscow (Russia) from 3rd till 7th October 2016 on the base of the Institute for Spectroscopy RAS.

The Congress program covered a broad range of actual questions of optics and spectroscopy, including fundamental atomic and molecular spectroscopy, spectroscopy of condensed matter and plasma, spectroscopy and microscopy of single quantum objects and nanostructures. Various interdisciplinary fields were also included and suggested for discussions during the meeting: quantum informatics, nanoplasmonics and metamaterials, nanooptics, photonics, applied spectroscopy in bio- and medical physics. Some special attention was paid to the questions of modern spectral instrumentation.

As a side event to the Congress there was the Youth Scientific School on Optics and Spectroscopy that offered a series of lectures and included talks of young scientists followed by a competition of their scientific works. The school was organized with help of the Moscow State Pedagogical University (www.mspu.ru).

The Congress was also accompanied with the exhibition organized by sponsoring companies, where the modern equipment for optics, spectroscopy and related areas from the world-wide leading manufacturers was demonstrated.

The scientific program of the meetings included 25 invited lectures, 54 oral talks, 112 poster talks, authored in total by 605 scientists from 16 countries (Russia, Belarus, Armenia, Ukraine, Italy, Germany, France, Austria, Poland, Czech Republic, Sweden, India, China, Japan, Australia, USA).

* Corresponding author: naumov@isan.troitsk.ru; web-page: www.spectroscopy.su

This issue of the EPJ Web of Conferences consists of the expanded version of abstracts of the selected talks presented at the Congress. The issue is recommended for specialists who work in the field optics and spectroscopy, as well as for M.S. and Ph.D. students. The full program and details about the Congress can be found on the web-site www.spectroscopy.su). The next 26th Congress will be organized by the Institute for Spectroscopy RAS in 2020.

The general photo of the 25th Congress on Spectroscopy.

Opening session (left). The organizing committee (right).

The financial support is acknowledged from the Federal Agency of Scientific Organisations of Russian Federation and from Russian Foundation of Basic Researches (16-02-20722 – support of the Congress; 16-32-10405 - support of Youth scientific school).

The local organising committee of the Congress is a part of the leading scientific school of Russia NSh-7035.2016.2 "Spectroscopy of atoms, molecules and condensed matter".

The meeting and school were also sponsored by several companies: "Avesta Project" (manufacturing of lasers and optical components, www.avesta.ru); "Azimuth Photonics" (distribution of optoelectronics components, devices and research equipment for the photonics industry, www.azimp.ru); "Fizmatlit" (the publishing house, www.fml.ru); "Scontel" (manufacturing of the single photon detectors, www.scontel.ru); "SOL Instruments" (manufacturing of technologically advanced instruments for spectroscopy and related fields, www.solinstruments.com); "YE-International" (distribution of electronics and electromechanical components, measuring devices, instruments, power supplies, tools and accessories, www.yeint.ru).

edp sciences

EPJ Web of Conferences

Statement of Peer review

In submitting conference proceedings to EPJ Web of Conferences, I certify to the Publisher that I adhere to the **Policy on Publishing Integrity** of the journal in order to safeguard good scientific practice in publishing.

1. All articles have been subjected to peer review administered by the proceedings editors.
2. Reviews have been conducted by expert referees, who have been requested to provide unbiased and constructive comments aimed, whenever possible, at improving the work.
3. Proceedings editors have taken all reasonable steps to ensure the quality of the materials they publish and their decision to accept or reject a paper for publication has been based only on the merits of the work and the relevance to the journal.

Title, date and place of the conference

XXV-th Congress on Spectroscopy
October 3 - 7, 2016
Moscow, Troitsk, Russia
Institute for Spectroscopy of the Russian Academy of Sciences

Proceedings editor(s):

Prof. Dr. Evgeny A. Vinogradov
Prof. Dr. Andrei V. Naumov
Dr. Maxim G. Gladush
Dr. Kamil R. Karimullin

Date and editor's signature

November 25, 2016

 DOI: 10.1051/epjconf/201713201001

Measurement of the parameters of non-stationary gas flows by diode laser absorption spectroscopy in case of high temperature and high pressure

M.A. Bolshov[1,*], V.V. Liger[1], Yu.A. Kuritsyn[1], V.R. Mironenko[1], Ya.Ya. Ponurovskii[2], and O.M. Kolesnikov[3]

[1]Institute for Spectroscopy RAS, 108840 Troitsk, Moscow, Russia
[2]A.M. Prokhorov General Physics Institute RAS, 119991 Moscow, Russia
[3]N.E. Zhukovsky Central Aerohydrodynamic Institute, 140180 Zhukovsky, Russia

Abstract. Experimental version of diode laser absorption spectrometer (DLAS) for contactless measurements of temperature and water vapor concentration in supersonic gas flows is developed. The spectrometer can be used for the measurements of temperature up to 2500 K and total pressure up to 3 atm. The technique is based on the registration of the transient absorption spectra of a target molecules and fitting of the experimental spectra by the simulated ones constructed using the spectroscopic databases. The temperature is inferred from the ratio of the intensities of the absorption lines with different low energy levels. In gas media with the above parameters the absorption lines are broadened which demands the use of two diode lasers (DL) working in different spectral ranges. The software for selection of the optimal line combinations was developed. The combination of two strong lines in the spectral ranges 1.39 μ and 1.34 μ was selected as the optimal one. The efficiency of the developed technique was exemplified in the first set of the experiments in conditions of real propulsion in Zhukovsky Central Aerohydrodynamic Institute (TsAGI) for the temperatures within (500-2200) K range and total pressure up to 3 atm.

Absorption spectroscopy with diode lasers (DLAS) is widely used for diagnostics of gases [1-6]. It is particulary the most popular technique for contactless probing of different combustion zones in mixing gas flows because any traditional thermocouples being inserted in a flow strongly modifies the structure of the flow.

General approach of DLAS technique is based on the measurement of the integral intensities of two lines of a test molecule with different energies of low levels. Assuming the termodynamic equilibrium the ratio of the line intensities *S(T)* depends only on gas kinetic temperature. In case of low pressure (<1 atm.) and temperature (<1500 K) the absorption lines are relatively narrow and can be spectrally resolved. One can find a spectral interval in near-ir with different absorption lines of water molecules lieing within a

* Corresponding author: mbolshov@mail.ru

range of fast tuning of a diode laser DL (2-3 cm^{-1}). For the diagnostics of such gas medium a single DL can be used.

The situation becomes more problematic if total gas pressure is above 1 atm. and temperature is above 2000 K. In case of high pressure and temperature the absorption lines are broadened and one cannot find a spectral interval of about 3 cm^{-1} (interval of a fast tuning of a DFB DL) with isolated lines. For such a case two DLs generating in different spectral ranges should be used. Evidently, the spectral intervals with dominating strong lines should be selected. For the above experimental parameters we accounted for the contribution from the neighboring lines and searched for spectral intervals of 3 cm^{-1} with water molecule lines with close low levels.

The criteria of line selection and different combinations of the absorption lines for DLAS technique are developed. The software for the selection of the lines from the data bases HITRAN-2012 and HITEMP satisfying the criteria of minimal errors of temperature measurement δT for the actual experimental error of integral line intensity measurement δS is developed. Different algorithms of data processing were checked in numerical experiment.

Specific units of the developed DLAS spectrometer and different algorithms of data processing of the trancient absorption spectra for final evaluation of temperature are described. DLAS enables to measure the temporal evolution of gas flow temperature and water vapor concentration with time resolution of about 1 ms. This resolution is mainly limited by the upper limit of the DL tuning frequency (10 kHz for DFB type lasers). Two optimal spectral intervals for the measurements of gas temperature at total pressure up to 3atm. and temperature around 2000 K were selected. The criteria for line selection and the attainable errors in temperature determination are discussed. The developed algorithm of temperature evaluation was tested in numerical experiments. The spectra in both selected intervals were simulated using spectral data bases for the expected parameters of a hot zone – total pressure up to 3 atm., temperature up to 2000 K and concentration of water vapor of several percents. Then white noise with different mean powers was added to the simulated spectra and these noisy («experimental») spectra were evaluated by different developed algorithms. It was proved that in case of large S/N ratio minimal errors in temperature evaluation can be realized if maximal number of absorption lines within the range of DL tuning with integral intensities above 10% of the most intense line are included in fitting.

The desined and produced new version of DLAS was tested on propulsion system in Zhukovskii Zentral Aerohydrodynamic Institute. In probed zone of the hot gas flow temperature in the range 600-2000 K and pressure 0.5 – 3 atm. were measured.

References

1. M.G. Allen, Meas. Sci. Technol. **9**, 545 (1998)
2. R.K. Hanson, Proc. Combust. Inst. **33**, 1 (2011)
3. M.A. Bolshov, Y.A. Kuritsyn, Y.V. Romanovskii, Spectrochim. Acta B **106**, 45 (2015)
4. C. Schulz, A. Dreizler, V. Ebert, J. Wolfrum, *Combustion Diagnostics. Springer Handbook of Experimental Fluid Mechanics* (Springer, Berlin, 2007)
5. Z.P. Wang, F. Li, H.B. Gu, X.L. Yu, X.Y. Zhang, Aerosp. Sci. Technol. **42**, 169 (2015)
6. M.A. Bolshov, Y.A. Kuritsyn, V.V. Liger, V.R. Mironenko, S.B. Leonov, D.A. Yarantsev, Appl. Phys. B **100**, 397 (2010)

EPJ Web of Conferences **132**, 01002 (2017) DOI: 10.1051/epjconf/201713201002
SPECTROSCOPY.SU 2016

IR surface polariton spectroscopy as a method for studying the optical properties of ultra thin films

E.A. Vinogradov*

Institute for spectroscopy RAS, 108840 Troitsk, Moscow, Russia

Abstract. Experimental results of investigation of optical properties of MgO thin films (thickness 10, 30, 100 и 300 nm) and AlN films (thickness 40 и 400 nm) on sapphire substrate are discussed. The optical phonon frequencies of these films are located in frequency region of surface polariton of sapphire. Due to the resonance between them the splitting and the shift of absorption spectra of sapphire surface polariton appear. From these experimental data it is possible to reconstruct all constants of the permittivity of both the film and substrate, the film thickness, and to specify its structure.

To use thin semiconductor films in a variety of applications, including optoelectronics, need to know their physical properties. In most cases the film properties are very different from the properties of bulk materials from which they are made. To study the physical (optical) properties of ultrafine (nanometer) films still no spectrometers required photometric accuracy. In some cases, to determine the physical parameters of nanofilms useful surface polariton spectroscopy.

Surface polaritons (SP) are nonradiative electromagnetic excitations at the interface of two media, propagating along the interface, if the dielectric constant of the contacting materials have different signs [1-4]. Electromagnetic field of SP focuses directly from the media interface, and decreases exponentially with distance from the interface (near field), so that the SP is very sensitive to the characteristics of the interface. If the frequency of the optical phonons of the film lies in the region of existence of SP substrate, then, by the resonance between them, there is a splitting of the bands and a shift of the absorption spectra of SP substrate. From these experimental data can be restored all constant permittivity and the thickness of the film, and thus to characterize the film properties. The resulting gap in this dispersion curves of SP substrate is proportional to the square root of the thickness of the film that allows you to measure the thickness of very thin films [1, 5-9]. The report discusses the experimental results on the optical properties of the films of MgO (thickness 10, 30, 100 and 300 nm) [5, 7, 9] and films of AlN (thickness of 40 nm and 400 nm) on sapphire [6-9]. SP spectra were measured in attenuated total internal reflection

* Corresponding author: evinogr@isan.troitsk.ru

mode (ATR) on the FT-IR spectrometer IFS66v (Bruker), and range of external (in the far field) reflection at angles close to normal incidence. The external reflection spectra of MgO film thickness of 10 nm and 30 are not visible, and the spectra of the film thickness of 100 nm and 300 nm appear only at low frequencies (near 400 cm^{-1}) due to resonance TO phonon of the film with low frequency TO phonon sapphire. By measuring the ATR spectra of film on sapphire at different angles of incidence of light in the prism ATR can restore dispersion SP sapphire with the film. MgO film radically changes the spectrum of sapphire SP due to the resonance of sapphire SP with an optical phonon of MgO film. This interaction leads to splitting of the high branches of SP sapphire near the frequency of 700 cm^{-1}. From the angular dependence of the reflectance spectra in ATR mode dispersion of SP sapphire with the film was reconstructed. It was compared with a dispersion of SP, calculated using the parameters of the dielectric constants of sapphire (found from the analysis of variance external reflection spectra) in the fitting parameters of the dielectric constant of the film, which determines the best agreement between the calculated variance of SP with experimental frequency-angular spectrum. Fig. 1 shows the dispersion curves of SP 10 nm MgO film on sapphire, which the spectra of the far-field was not visible at all.

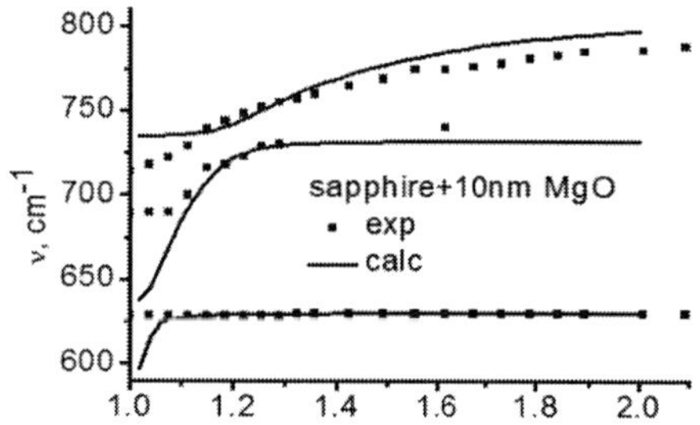

Fig. 1. The dispersion of surface polaritons of "MgO film 10 nm thick on sapphire." The solid curves are calculated using the optical constants of bulk MgO [5], and sapphire.

Surprisingly, studies have shown that the optical constants of MgO film thickness of 10 and 300 nm were very close to those of bulk MgO crystals, in contrast to the film thickness of 30 and 100 nm.

Thus, the example of MgO films on sapphire shows that due to the resonance of SP sapphire substrate with optical phonons even very thin (nm) film, it is possible to restore the dispersion curves of the surface polaritons.

As a result, the calculated dispersion curves fit to the experimental, varying the dielectric constant of the film permeability, it can be determined not only all these constants, but also to receive information about modifying the optical constants of the substrate material on the nanometer thickness [5-9].

Support from the Grant of the President of Russia for Leading Scientific Schools # 7035.2016.2 is acknowledged.

References

1. G.N. Zhizhin, M.A. Moskaleva, E.A. Vinogradov, V.A. Yakovlev, Appl. Spectr. Rev. **18**, 171 (1982)
2. E.A. Vinogradov, I.A. Dorofeev. *Thermally stimulated electromagnetic fields of solids* (Fizmatlit, Moscow, 2010)
3. I.A. Dorofeyev, E.A. Vinogradov, Phys. Rep. **504**, 75 (2011)
4. I.A. Dorofeyev, E.A. Vinogradov, Int. Rev. Phys. **6**, 202 (2012)
5. V.A. Yakovlev et al., J. Nanopart. Res. **13**, 5841 (2011)
6. V.A. Yakovlev, N.N. Novikova, E.A. Vinogradov, S.S. Ng, Z. Hassan, H. Abu Hassan, Phys. Lett. A **373**, 2328 (2009)
7. E.A. Vinogradov, N.N. Novikova, V.A. Yakovlev, Physics – Uspekhi **184**, 653 (2014)
8. N.N. Novikova et al., Surface & Coatings Technology **227**, 58 (2013)
9. N.N. Novikova, V.A. Yakovlev, E.A. Vinogradov, S.S. Ng, Z. Hassan, H.A. Hassan, Appl. Surf. Sci. **267**, 93 (2013)

Increasing the efficiency of organic solar cells using dielectric spherical nanoparticles

Yu.V. Vladimirova[1,*], and V.N. Zadkov[1,2,3]

[1]International Laser Center and Faculty of Physics, Lomonosov State University, 119991 Moscow, Russia
[2]Kolmogorov School, Lomonosov State University, 119991 Moscow, Russia
[3]Institute for Spectroscopy RAS, 108840 Troitsk, Moscow, Russia

Abstract. In this paper we review recent progress in using plasmonic nanoparticles for improving efficiency of the organic photovoltaic (PV) cells with embedded plasmonic nanoparticles. Specifically, we discuss how the plasmonic nanoparticles can be used for guiding and concentrating the light for enhanced absorption, paying attention to both fundamentals and design considerations, as well as to realization of the broadband plasmonic scattering to better utilize the incoming solar spectrum. Plus to that, we discuss potential advantages of using dielectric nanoparticles in PV solar cells.

The development of alternative sources of energy such as solar photovoltaic (PV) devices is one of the urgent problems of our society. In this field primary tasks are the increase of the solar cells efficiency and significant reduction in their price. For the crystalline silicon solar cells, the problem of increasing their efficacy was largely solved, however they are still very expensive. By contrast, organic solar cells are rather cheap in production, but have quite low efficiency so far [1]. That is why the development of various methods that allow to increase the organic solar cells efficiency is of prime importance [2].

One option to raise an efficiency of a PV solar cell is the use of plasmonic nanoparticles, which are embedded into the solar cell. Each nanoparticle serves then as a nanoantenna concentrating the incident electromagnetic field in the absorption area and thereby increasing the efficiency of the solar cell [3]. Such nanostructures are already used in silicon solar cells, but technologies of the nanostructure deposition are very expensive and laborious.

We discuss here the potential advantages of using dielectric nanoparticles in organic PV cells. High-refractive index dielectric nanoparticles may exhibit strong directional forward light scattering at visible and near-infrared wavelengths due to interference of simultaneously excited electric and magnetic dipole resonances [4,5]. For a spherical particle shape, the so-called first Kerker's condition can be realized, at which the backward scattering practically vanishes for some relation between the refractive index and the particle's size.

* Corresponding author: yu.vladimirova@physics.msu.ru

Experimentally, directional light scattering by spherical silicon nanoparticles in the visible spectral range was demonstrated, for example, in Ref. [6]. Many investigators considered scattering properties of a silicon nanoparticle in free space, but in the real applications the nanoparticles are often located near a substrate and the influence of the substrate may be very significant [5-7]. In this work, we derived optimal conditions for forward light scattering by spherical homogeneous dielectric nanoparticles with high refractive index in non-magnetic and non-absorbing media near a dielectric substrate.

We consider substrates with n=1.5 (P3HT:PCBM). Our calculations of the scattering diagram of a nanoparticle located near a dielectric substrate demonstrate its strong anisotropy in both forward and backward directions. We studied in detail how the scattering diagram of a spherical silicon nanoparticle near a dielectric substrate depends on the radius of the nanoparticle and distance to the substrate, as well as on the polarization, wavelength and angle of the incident light wave [8].

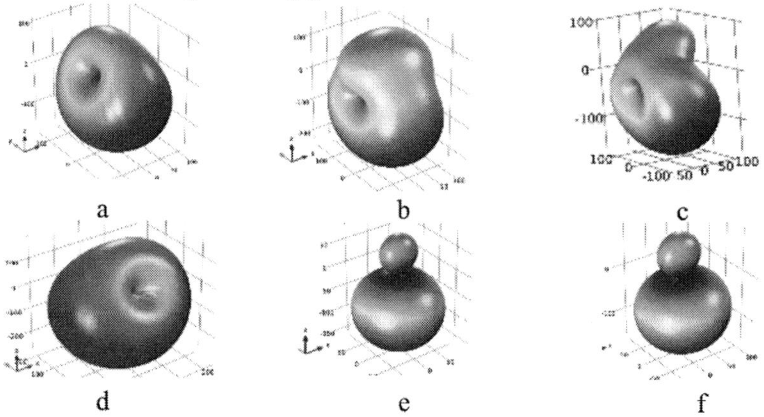

Fig. 1. Scattering diagrams of a silicon nanoparticle with a radius of 65 nm in free space (a,d and near a substrate (n=1.5) for normal incidence θ=0 (b,e) and $\theta=\pi/6$ (c,f) for the incidence radiation wavelength λ=550nm (a,b,c) and 650 nm (d,e,f).

Scattering diagrams of a silicon nanoparticle with a radius of 65 nm in free space (a,d) and near a substrate (n=1.5) for normal incidence θ=0 (b,e) and $\theta=\pi/6$ (c,f) for the incidence radiation wavelength λ=550 nm (a,b,c) and 650 nm (d,e,f) is presented in the figure. From this figure it follows that the maximum forward scattering for such silicon nanoparticle is observed for λ = 350, 460, 540, 640 nm for normal incidence $\theta = 0$ and $\theta = \pi/6$.

The authors would like to acknowledge financial support from the Russian Foundation for Basic Research under the grant No. 16-02-00816.

References

1. M. Gu, Z. Ouyang, J. Baohua et al., Nanophotonics **1**, 235 (2012)
2. Z. Tang, W. Tress, O. Inganäs, Materials today **17**, 389 (2014)
3. L. Novotny, B. Hecht, *Principles of Nano-Optics*, (Cambridge University Press, 2012).
4. Y.H. Fu, A.I. Kuznetsov, A.E. Miroshnichenko et al., Nat. Commun. **4**, 1527 (2013)
5. M.I. Tribelsky, J.-M. Geffrin, A. Litman, Ch. Eyraud, F. Moreno, Sci. Reports **5**, 12288 (2015)
6. A. Pors, S.K.H. Andersen, S.I. Bozhevolnyi, Opt. Express **23**, 28808 (2015)
7. A.E. Miroshnichenko, A.B. Evlyukhin, Y.S. Kivshar, B.N. Chichkov, ACS Photonics **2**, 1423 (2015)

Development of the experimental setup for investigation of latching of superconducting single-photon detector caused by blinding attack on the quantum key distribution system

M.S. Elezov[1,*], R.V. Ozhegov[1], G.N. Goltsman[1], and V. Makarov[2]

[1]Moscow State Pedagogical University, 119991 Moscow, Russia
[2]Institute for Quantum Computing, University of Waterloo, N2L 3G1 Waterloo, Canada

Abstract. Recently bright-light control of the SSPD has been demonstrated. This attack employed a "backdoor" in the detector biasing scheme. Under bright-light illumination, SSPD becomes resistive and remains "latched" in the resistive state even when the light is switched off. While the SSPD is latched, Eve can simulate SSPD single-photon response by sending strong light pulses, thus deceiving Bob. We developed the experimental setup for investigation of a dependence on latching threshold of SSPD on optical pulse length and peak power. By knowing latching threshold it is possible to understand essential requirements for development countermeasures against blinding attack on quantum key distribution system with SSPDs.

As is predicted by laws of the quantum mechanics, one can transmit information from a party A ("Alice") to a party B ("Bob") securely without successful interception by a third party ("Eve"), as such an interception would be immediately detected as absence of the expected photon or inconsistency of its state [1]. Both single-photon Avalanche PhotoDiodes (APDs) and Superconducting Single-Photon Detectors (SSPDs) [2] are employed in commercial Quantum Key Distribution (QKD) systems [3].

SSPD outperforms APD by their characteristics such as low dark counts, high quantum efficiency at wavelength of 1550 nm and low jitter, which allows for increasing secret key rate, distance of quantum key distribution and for decreasing the quantum bit error rate (QBER) [4]. At the same time, development of quantum cryptography systems and their emerging to a commercial level stimulates research of potential vulnerabilities for hacking attack. One of many potential points for hacking is the superconducting single-photon detector. Some maleficent ways for remote control of an APD [5] and SSPD [6, 7] by blinding attack were demonstrated. In the blinding attack, the detector is subjected to a strong optical pulse leading to the detector being in a non-operational state for some time.

The potential vulnerability of the SSPD is the possibility of latching. After absorption of a strong optical pulse, the SSPD switches to normal state and is heated up by the bias current. The SSPD temperature is increased up to the resistive transition temperature, and the resistance of the SSPD rises up to 1 - 2 MOhm. Further increase of the SSPD

* Corresponding author: elezovms@rplab.ru

temperature occurs due to the Joule heating. The equilibrium temperature is reached when equality between the heat power and dissipated power is achieved. The SSPD stays in the normal state without imminent return to superconducting state. Such phenomenon is called latching [8], which can be used by Eve in the blinding attack. For that Eve blocks signals from Alice, blinds Bob's SSPDs by radiation at wavelength of 1550 nm and forces Bob's readout. We developed the experimental setup which is needed for investigation of latching threshold of SSPDs.

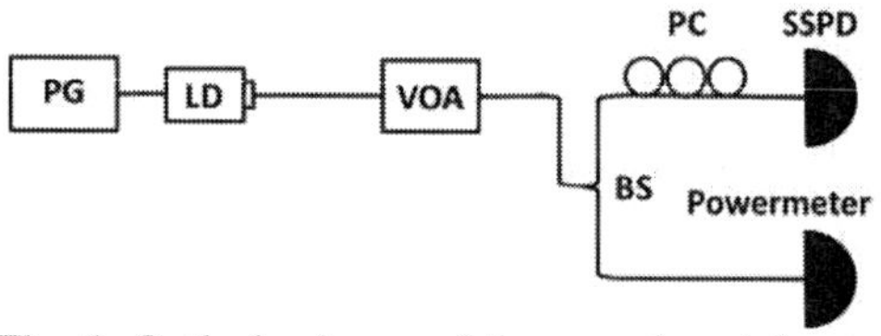

Fig. 1. Optical scheme of the experimental setup. PG – pattern generator; LD – laser diode at 1550 nm; VOA – variable optical attenuator; BS – beamsplitter 50/50; PC – polarization controller.

For measurement of the latching threshold of the SSPD we assembled the experimental setup shown in Fig. 1. In our experiment we use one SSPD of registration system. It is designed as a cryogen-free system and involves a compact low-power cold head SRDK-101D with cryostat, including a closed-cycle refrigerator, a compressor with helium lines and an SSPD control unit. The operating temperature is 2.5 K. An advantage of such a system is that closed-cycle refrigerators can operate continuously for months provided there is a reliable power source. Such a cryostat has two SMA outputs for biasing the SSPDs and reading out the output signal. Besides, the cryostat has two optical FC/PC inputs. Optical coupling is realized by means of the alignment of the 9 μm single-mode fiber core and the meander of the SSPD measuring 7×7 μm^2. The size of a meander is optimized with respect to optical matching with the core of a single-mode fiber SMF-28e. After the final alignment, the optical unit was mechanically fixed on the cold finger of the cryostat. The SSPD operates in a single-photon mode with high quantum efficiency above 20%, low dark count rate and relative bias current of about $0.95I_c$. The latching threshold of the SSPD depends on the peak power, pulse length and pulse repetition rate of the strong optical pulses. The laser diode is controlled by the pattern generator. The laser light with linear polarization at 1550 nm is sent to the beamsplitter 50/50. By means of the power meter and the variable optical fiber attenuator we will measure peak power in optical pulses incident on the SSPD. The polarization controller is needed for tuning light polarization so that the vector of polarization of the light is parallel nanowires, as it ensures the maximal quantum efficiency of the detector.

Thus, we developed the experimental setup for investigation of the SSPD latching threshold dependence on the optical pulse length and its peak power. By knowing conditions for the SSPD latching one could develop countermeasures against blinding attack on quantum key distribution system based on the superconducting single-photon detectors.

The work was supported by The Ministry of Education and Science of Russian Federation, project No. 14.586.21.0007, RFMEFI58614X0007.

References

1. N. Gisin et al., Quantum Cryptography II Rev. of Mod. Phys. **74**, 145 (2002)
2. G. Gol'tsman et al., Appl. Phys. Lett. **79**, 705 (2001)
3. http://www.idquantique.com
4. R. Ozhegov et al., Proc. of SPIE **9440**, 94401F (2014)
5. L. Lydersen et al., Nature Photonics **4**, 686 (2010)
6. L. Lydersen et al., New J. Phys. **13**, 113042 (2011)
7. M.G. Tanner et al., Opt. Express **22**, 6734 (2014)
8. A. Annunziata et al., J. Appl. Phys. **108**, 084507 (2010)

Heterodyne spectroscopy with superconducting single-photon detector

Yu.V. Lobanov[1,2,*], M.L. Shcherbatenko[1,2], A.V. Semenov[1,2], V.V. Kovalyuk[1], A.A. Korneev[1,2], and G.N. Goltsman[2,3]

[1]Moscow Institute of Physics and Technology, 147000 Dolgoprudny, Russia
[2]Moscow State Pedagogical University, 119991 Moscow, Russia
[3]National Research University Higher School of Economics, 101000 Moscow, Russia

Abstract. We demonstrate successful operation of a Superconducting Single Photon Detector (SSPD) as the core element in a heterodyne receiver. Irradiating the SSPD by both a local oscillator power and signal power simultaneously, we observed beat signal at the intermediate frequency of a few MHz. Gain bandwidth was found to coincide with the detector single pulse width, where the latter depends on the detector kinetic inductance, determined by the superconducting nanowire length.

Heterodyne detection is one of the most powerful spectroscopic tools. Although it originated from the radio-frequency band, its applicability in optics has been suggested and demonstrated [1-2]. With invention of the laser, use of the heterodyne technique in optics has been increased drastically, including various applications in astrophysics, spectroscopy, lidar development, communication. With aim to improve performance of the receivers, use of photon-counting detectors has been proposed [3], and APDs array has been utilized for the lidar experiment [4]. Here we demonstrate successful operation of a Superconducting Single Photon Detector (SSPD) [5,6] as the core element for heterodyne spectroscopy.

We used two independent fibre-coupled DBF-lasers operating at wavelength near 1550 nm as the local oscillator (LO) and signal source. The wavelength of each laser was precisely set by adjusting its operating temperature through a computer-controlled bias unit. The output power from each laser was adjusted by optical attenuators and the polarization was adjusted by the polarization controllers. Then waveforms from LO and signal were combined by the beamsplitter and fed to the SSPD operated in liquid helium cryostat at 1.7 K temperature. The photon detection events - several nanoseconds-long electrical voltage pulses - were amplified by two room temperature amplifiers with 1 GHz band and total gain of about 50 dB and then fed to either digital oscilloscope Rohde & Schwartz RTO 1022 or Spectrum Analyzer R&S FSV 13. We stored sub-millisecond train of pulses in the oscilloscope memory and then performed their statistical analysis.

When both waveforms are combined the power of the resulting field experiences beating at an intermediate frequency (IF) which is the difference between LO and signal frequencies. Consequently, the probability of detection event exhibits the same beating at IF. Fig.1 (a) demonstrates a zoomed-in view on the train of pulses recorded with the SSPD. Given the

* Corresponding author: akorneev@rplab.ru

 DOI: 10.1051/epjconf/201713201005

recorded set of detection events one can then digitally perform Fourier-transform analysis and find IF. Fig. 1 (b) presents the result of such analysis, clearly showing IF signal at ~11MHz fitted by the Cauchy-Lorentzian distribution.

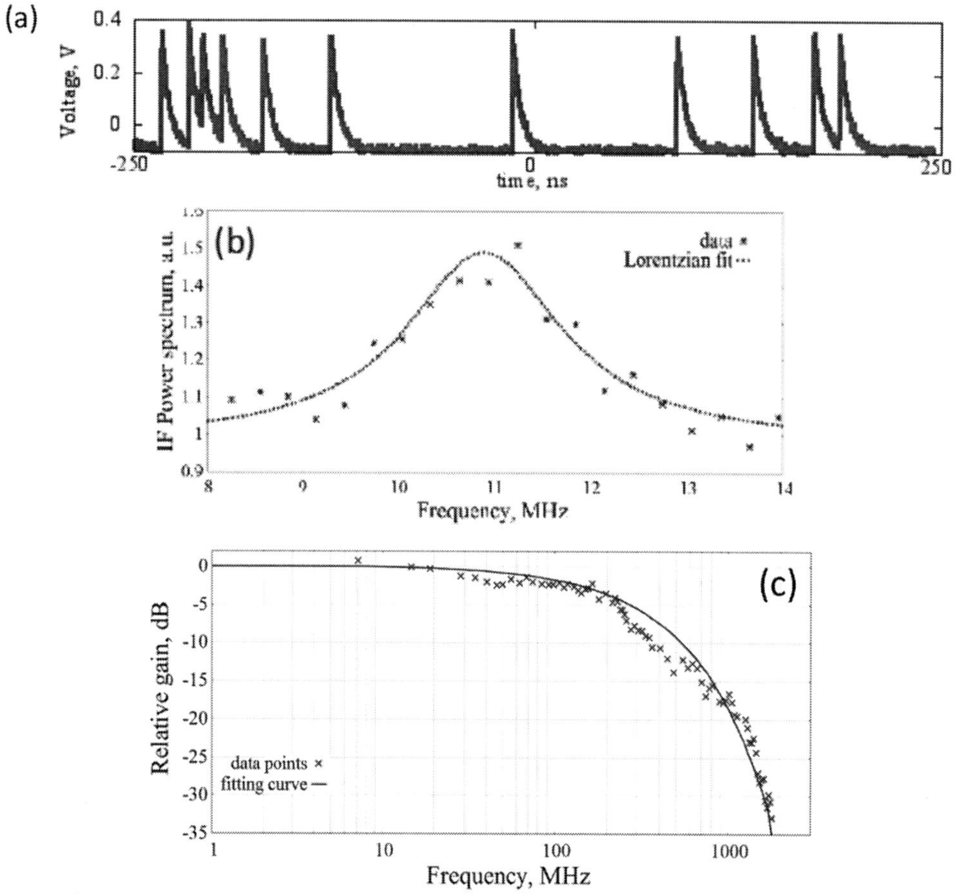

Fig. 1. (a) Trace of photon detection voltage pulses distributed with the beating frequency. (b) Result of Fourier analysis of detection events distribution in time presenting maximum at the intermediate frequency. (c) Gain bandwidth of the SSPD operated in heterodyne detection regime.

One of the important characteristics of a mixer is the mixer gain bandwidth. With the analog spectrum analyzer we measured the output power at the IF. By changing the local oscillator wavelength, we were able to determine the gain bandwidth of our system, as shown in Fig.1 (c). The measured gain bandwidth was as large as 140 MHz.

This work was supported by Russian Science Foundation project 16-19-10633.

References

1. A.T. Forrester et al., Phys. Rev. **72**, 728 (1947)
2. A.T. Forrester, R.A. Gudioindsen, P.O. Johnson, Phys. Rev. **99**, 1691 (1955)
3. D.E. Sawyer, *Heterodyne operation of avalanche photodiodes* (NASA, USA, 1967)
4. J.X. Luu, L.A. Jiang, Appl. Opt. **45**, 3798 (2006)
5. G. Gol'tsman et al., Appl. Phys. Lett. **79**, 705 (2001)
6. C.M. Natarajan et al., Supercond. Sci. Technol. **25**, 063001 (2012)

Technical and technological investigation of cultural heritage: experience in applications of spectroscopic methods

I.A. Grigorieva[1,*], K.S. Chugunova[1], I.F. Kadikova[2], S.V. Khavrin[1], and S.A. Pisareva[2]

[1]The State Hermitage Museum, 190000 Saint Petersburg, Russia
[2]The State Research Institute for Restoration, 107014 Moscow, Russia

Abstract. Physical and chemical methods of analysis are indispensable for restoration, study of painting techniques, examination and attribution of works of art. Development of new directions of investigations as well as improvements in sample preparation allow applying non-destructive analysis methods, minimizing amount of matter used to obtain informative spectra, preventing alteration and destruction of samples in the course of investigation. This paper observes the examples of applying optical and spectral methods, including molecular spectral analysis and electron microscopy, for study of binding media and pigments of painting and archaeological artifacts.

Preservation of cultural heritage, study and evaluation of their condition are the main purposes of restoration and research. Therefore, analysis of works of art is realized by means of non-destructive methods, including investigation in visible, infrared, ultraviolet and X-ray light as well as optical microscopy and XRF [1].

The first stage of research involves non-sampling methods, for example, XRF and reflectance spectroscopy – which in some cases are the only possible way to examine works of art such as paper, fabrics, aquarelles and manuscripts. These methods provide the information regarding just the superficial layer. Applying them during investigation of K. Bryullov's drawings from "Italian Album" series apart from frequently used Prussian blue it was identified rare organic pigment Indian yellow (mixture of calcium and magnesium salt of euxantic acid). Indian yellow was used in Europe from late XVIII century till 1908 when its production was restricted [2]; it corresponds to period of the artist' life.

Detailed information about multicomponent multilayer structures (painting, ceramics, archaeological objects etc.) can be obtained only by taking samples and their investigation by a combination of analytical methods. Thus, lead white in the form of unusually large tabular hexahedral crystals was discovered in pictures of K. Malevich by means of polarized light microscopy [3]. SEM-EDX method confirmed lead presence in crystals and determined their size (about 7 μm), and XRD showed that these crystals are composed of hydrocerussite ($2PbCO_3 \cdot Pb(OH)_2$).

* Corresponding author: grigorieva_ia@hermitage.ru

Sampling is another key issue in obtaining relevant results. Thus, micro-ATR and micro-Raman spectroscopy were chosen as the most suitable methods for study of molecular composition of degraded historic blue-green glass beads because of their dimensions, geometry peculiarities and necessity of their preservation. The main problem was connected with beads fixation on microscope stage and was solved using an Indium plate in which the sample was placed.

Another issue which complicates investigation and interpretation of results is connected with superposed composition of samples. Most museum objects consist of both organic and inorganic compounds, including natural ones with variational chemical composition which are essentially altered due to ageing and restoration interventions. Long-time modeling and artificial ageing don't always lead to understanding of physical and chemical processes which take place for a long period of time and consequent extrapolation doesn't provide reliable estimation of results [4, 5], for example, age determination of works of art by means of FTIR-spectroscopy.

Multidisciplinary investigation of Russian oil painting of XX century based on lead white and zinc white was carried out. Analysis of calibration dependences of integral intensities of absorption bands at 1740 cm^{-1} (carbonyl peak) and 1650-1500 cm^{-1} region (metal carboxylates peaks) on time showed their substantially non-linear behavior [6]. But in course of study a number of technological features of paints preparation was revealed; these features can be used as time markers. For example, it was typical for oil paints based on zinc white of early XX century (up to 30s) the presence of gypsum (from 10 to 50%), barium white (up to 20%) and lead white (up to 10%), and, less commonly, china clay, chalk and talc.

The paper shows the necessity of development of complex approach and strategy for investigation of cultural heritage objects. This allows minimizing negative impact on works of art and obtaining relevant results for different museum tasks.

The support from Russian Foundation for Basic Research is acknowledged (14-06-00244).

The support from Russian Science Foundation is acknowledged (16-18-10366).

References

1. J.P. Mohen, M. Menu, B. Mottin, *Mona Lisa: Inside the painting,* (Abrams, New York, 2006)
2. N.S. Baer, A. Joel, R. L. Feller, N. Indictor, *Artists Pigments: A Handbook of their History and Characteristics* (Cambridge University Press, Oxford, 1986)
3. S. Pisareva, *XV Scientific Conference Expertise and Attribution of Works of Fine Arts* (Moscow, 2014)
4. J. van der Weerd, R.M.A. Heeren, Stud. Conserv. **49**, 193 (2004)
5. C. Miliani, F. Rosi, A. Daveri, Appl. Phys. A **106**, 295 (2012)
6. I.A. Balahnina., N.N. Brandt, A.Y. Chikishev, Y.I. Grenberg, I.A. Grigorieva, I.F. Kadikova, S.A. Pisareva, Appl. Spectr. **70**, 1150 (2016)

Study of wide band-gap crystal $LiCaAlF_6$ by IR-reflection spectroscopy and *ab initio* calculations

N.N. Novikova, S.A. Klimin, B.N. Mavrin*

Institute for Spectroscopy RAS, 108840 Troitsk, Moscow, Russia

Abstract. Polarized IR-reflection spectra and results of *ab initio* calculations of vibrational and electronic properties of $LiCaAlF_6$ single crystal are presented. It is shown that the crystal band gap is direct. Experimental and theoretical parameters are obtained for dipole-active and all phonons, respectively, including silent modes. Experimental IR-reflection and Raman spectra are well described in the frame of results obtained by *ab initio* calculations. The peculiarities are discussed concerning the structure of electronic bands, the interatomic interactions, the character of lattice vibrations, and the phonon dispersion.

Isostructural crystals with general chemical formula $LiMM'F_6$ (M = Ca,Ba,Sr; M′ = Al, Ga) possess a very wide band gap (Eg = 12.65 eV for $LiCaAlF_6$ [1]). That's why they are perspective for applications in UV spectroscopic region. The crystal $LiCaAlF_6$, pure as well as doped by d- and f-ions, is intensively studied. The list of its applications includes laser emission, scintillators, visualization of neutrons, UV optical windows, etc. The phonon spectrum of $LiCaAlF_6$, necessary for the description of relaxation processes and vibronic spectra of f-d transitions [2], is not sufficiently studied. Raman phonons were studied in Ref. [3]. In this work, we present results of IR spectroscopic study as well as of *ab initio* calculations of $LiCaAlF_6$.

Crystal structure of $LiCaAlF_6$ is trigonal [4], Li_2ZrF_6 type, space group is $P\bar{3}1c$ (D^2_{3d}, No. 163), z=2. Irreducable representations of phonon modes are classified as following $\Gamma^{vib}=3A_{1g}+5A_{2g}+8E_g+4A_{1u}+6A_{2u}+10E_u$ [3], A_{1g} and E_g being Raman active, A_{2u} and E_u – IR-active, A_{2g} and A_{1u} – silent modes.

Reflectivity spectra were registered using Brucker IFS66 fourier-spectrometer. *Ab initio* calculations were performed by two methods. Phonon properties in the center of Brillouin zone were calculated in the DFT-LCAO approximation with the use of CRYSTAL 14 code [5] and B3LYP hybrid pseudopotential [6]. Calculations of phonon and electronic dispersions across Brillouin zone were performed in the basis of plane waves using the code QUANTUM ESPRESSO [7].

Experimental reflectivity spectra of $LiCaAlF_6$ are presented in Fig.1 by symbols. The experimental parameters of IR phonons were obtained by fitting the spectra in the code RefFIT [8] using Drude-Lorentz model for dielectric constant. Model spectra with the best fit are presented for comparison in Fig. 1 by grey curves. The following TO frequencies (cm^{-1}) for IR active modes were obtained: 131, 284, 299, 400, 588 (A_{2u} modes), 122, 190,

* Corresponding author: mavrin@isan.troitsk.ru

 DOI: 10.1051/epjconf/201713201007

197, 269, 324, 360, 417, 451, 575 (E_u). We have to note that weak phonon modes (marked by arrows in Fig. 1) were reliably identified after *ab initio* calculation performed.

Calculated by both *ab initio* methods structural parameters coincide with experimental ones [9] within 1 % accuracy. Partial densities of electronic states and dispersion of electronic states was obtained, according to which, $LiCaAlF_6$ is a direct band-gap isolator. Theoretically obtained phonon frequencies are in good agreement with experimental ones, both IR and Raman active. The energies (in cm^{-1}) of silent phonon modes were also obtained (LCAO gives 244, 275, 406, 579 (A_{1u} modes), 31, 237, 305, 488, 726 (A_{2g})). Calculated IR reflectivity spectra (black curves at the bottom of Fig. 1) describe well experimental results. Analysis of TO-LO splittings was performed. One inverted phonon was found (E_u mode 451 cm^{-1}). Due to the interaction of weak phonon, falling into the TO-LO gap of the strong mode, with the last one, its TO-LO splitting inverts [10-13].

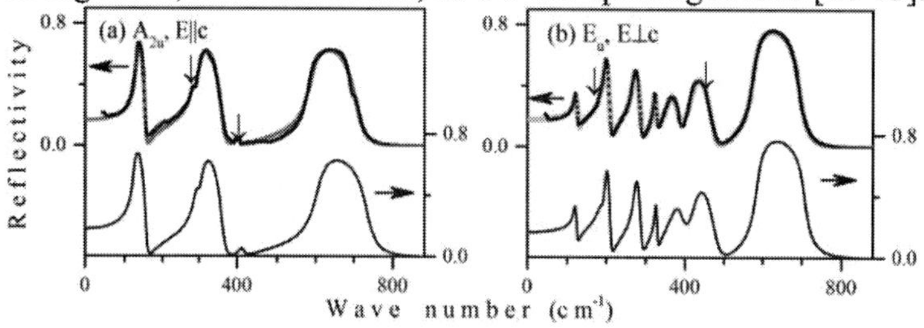

Fig. 1. Experimental (symbols), model (grey curves), and calculated *ab initio* (black curves at the bottom) reflectivity spectra of $LiCaAlF_6$ for (a) A_{2u} and (b) E_u vibrational modes. Arrows point to weak phonons

The phonons with energies higher than 500 cm^{-1} are separated by a gap from low-frequency ones. The analysis of atomic displacement in normal modes revealed predominantly stretching character of vibrations for high-frequency gapped modes. Phonon density of states calculated is also characterized by a gap in the 470-540 cm^{-1} spectral region. Phonon dispersion curves demonstrates strong softening (up to 3-5 cm^{-1}), close to A-point of the Brillouin zone, for two phonon modes, namely, for the lowest-energy phonons with symmetry A_{2g} (98 cm^{-1}) and E_g (132 cm^{-1}). This can be interpreted as quasi-doubling of primitive cell along z axis.

The authors are grateful to S.L. Korableva, who provided us with the $LiCaAlF_6$ crystal.

References

1. M. Kirm et al., Nucl. Instrum. Methods Phys. Res. A **537**, 291 (2005)
2. V.N. Makhov et al., J. Lumin. **132**, 418 (2012)
3. P. Daniel, J.Y. Gesland, M. Rousseau, J. Raman Spectrosc. **23**, 197 (1992)
4. G. Brunton, Acta Crystallogr. B **29**, 2294 (1973)
5. R. Dovesi et al., Int. J. Quantum Chem. **114**, 1287 (2014)
6. A.D. Becke, J. Chem. Phys. **98**, 5648 (1993)
7. http://www.quantum-espresso.org
8. A.B. Kuzmenko, Rev. Sci. Instr. **76**, 083108 (2005)
9. A.C. Cefalas et al., Proc. SPIE **4766**, 171 (2002)
10. F. Gervais, Opt. Comm. **22**, 116 (1977)
11. E.A. Vinogradov et al., Phys. Usp. **52**, 290 (2009)
12. S.A. Klimin et al., Phys. Rev. B **93**, 054304 (2016)
13. T.A. Ivanova, B.N. Mavrin, Opt. Spectrosc. **119**, 816 (2015)

Nanolocalized UV source of laser radiation

P.N. Melentiev[1,*], A.E. Afanasiev[1], A.A. Kuzin[1,2], and V.I. Balykin[1]

[1]Institute for Spectroscopy RAS, 108840 Moscow, Troitsk, Russia
[2]Moscow Institute of Physics and Technology, 141700 Dolgoprudny, Russia

Abstract. We show the results of optical nonlinear interaction of laser light with plasmonic nanostructures. Creation of a nanolocalized source of UV radiation (λ = 260 nm) of the record high intensity was demonstated. Possibility of use of the source for biomedical applications was shown.

Basic elements of nanoplasmonics are formed by metal structures having nano meter-scale geometry. Interaction of nanostructures with laser light at the frequencies of the localized plasmonic resonances leads to a high values of electric field near the nanostructure more than the amplitude of the incident light wave. Thus, nanostructure helps to localize light energy on a size considerably smaller compared to the wavelength [1]. There are different practical applications of the local field enhancement known such as optical microscopy with nano-meter resolution, biosensors, optical nanolithography, photovoltaics, light registration and photodetectors.

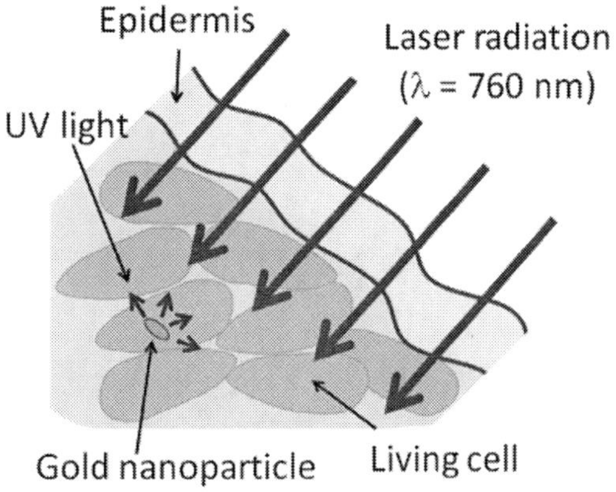

Fig. 1. Schematic figure of the application of the strong optical nonlinearity of plasmonic nanostructures for the UV treatment of living cell: UV radiation is formed by plasmonic nanostructure as a result of the harmonics generation. Radiation at fundamental frequency has the wavelength corresponding to a maximum value of the transmittance spectral window of a biological tissue.

* Corresponding author: melentiev@isan.troitsk.ru

A separate practical interest is a generation of harmonics by nanostructures in the UV spectral range. First of all this helps to realize nanolocalized sources of UV radiation, needed for nano-diagnostics and nanolithography. As a second, high optical nonlinearity of nanostructures in the UV spectral range can be used in the photo induced therapy of living cells (Fig. 1). Generation of a radiation in the UV spectral range by plasmonics nanostructures are not studied well. There are currently only several experimental papers where was demonstrated generation of third harmonics in the UV spectral range.

In this talk we present our results of optical nonlinear interaction of a femtosecond laser radiation having duration of 6 fs and 60 fs with individual plasmonic nanostructures made in aluminum and gold nanofilms. Chosen geometry of nanostructures helps to realize plasmonic resonance at the frequency of laser radiation, that leads to generation of the third harmonics in the deep UV spectral range, with a wavelength equals to 260 nm. We show that nanostructures made in gold nanofilm can be used for building the nanolocalized sources of UV radiation with a record high intensity.

This work was supported by the Russian Science Foundation (project No.14-12-00729).

References

1. L. Novotny, N. van Hulst, Nature Photonics **5**, 83 (2011)
2. P. Sasanpour, B. Rashidian, B. Rashidian, M. Vossoughi, Nano **5**, 325 (2010)

Spectroscopy of single organic dye-molecules and semiconductor quantum dots: basic aspects and applications in nanoscopy

A.V. Naumov*

Institute for Spectroscopy RAS, 108840 Moscow, Troitsk, Russia
Moscow State Pedagogical University, 119991 Moscow, Russia

Abstract.

Spectroscopy and microscopy of single quantum emitters is a research direction that has become by now one of the most topical issues in physics, physical chemistry, biophysics and related sciences [1-4]. This direction was marked by particularly intensive development when some new artificial chromophores became available. Those were molecular complexes, organometallic and hybrid light emitters, dielectric nanoparticles and colloidal semiconductor nanocrystals (quantum dots, QDs). Besides, the man-created nano-objects could have the specified photophysical characteristics.

Registration of fluorescence images of single point emitters in the scheme using a scanning confocal or widefield microscope can give the source coordinates with a subdiffraction (nanometer) accuracy. [5] The reconstruction of the emitters' coordinates is performed from the analysis of the point spread function (PSF). PSF can be modified to implement the super-resolution of all three coordinates using the adaptive optics elements, e.g., the double helix scheme [6]. In this case, the point source image becomes a complicated double-lobe shape and requires a specific analysis.

Restoration of coordinates for a statistically large number of single probe chromophores allows one to reconstruct the structure of the material under study with nanometer spatial resolution. This principle is a basis for rapidly growing multidisciplinary area known as the far-field optical microscopy of ultrahigh resolution (nanoscopy) with the reconstruction of fluorescence images of single point emitters. [2]

Especially informative and the most sensitive to structural and dynamic parameters of the nearby surrounding are the zero-phonon spectral lines (ZPL) of single impurity organic molecules (SM). [7] ZPLs are extremely sensitive to static and dynamic local fields as well as guest-host interactions. [8-12] The principal task in the development of this direction is to find the regularities connecting microscopic information on the level of individual emitters and their nano-surrounding with the macroscopic characteristics of the bulk samples. [3] The way to solve this problem consists in the statistical analysis of data for a large number of SMs. [4]

* Corresponding author: naumov@isan.troitsk.ru; web-page: www.single-molecule.ru

Our team is involved in development of the hardware and software solutions and methods for rapid detection and computer analysis of fluorescence images, spectra and spectral trajectories for the gigantic number of single emitters. The developed techniques allow us to perform qualitatively new type of measurements in the field of spectroscopy of the impurity center which is registration of individual spectra and the spatial coordinates of effectively emitting centers in the macroscopic volume of the sample doped with fluorescent probe molecules [3, 4, 13, 14]. In particular, we have been carried out experiments on the far-field visualizing of local structural features in solid films (defects, cracks) with characteristic dimensions of a few nanometers.

The method is based on a series-parallel detection of fluorescence images and spectra of a giant amounts of single fluorescent molecules and statistical analysis of the data. SM emission is excited selectively by a narrowband laser frequency tuning to the resonance frequencies of SM electronic transitions. Analysis of SM spatial placement makes it possible to visualize the features of the solid structure of the sample with nanometer accuracy, exceeding the diffraction limit.

The features of nanoscopy expand qualitatively the detection of spatial and spectral (photophysical) characteristics of probe emitters. In fact, in these experiments, microscopy acquires extra "spectral coordinates", which are individual spectral characteristics of nanoprobes, which makes it possible to carry put a kind of "spectral nanotomography", revealing different characteristics of the local structure and dynamics of the sample (Fig. 1).

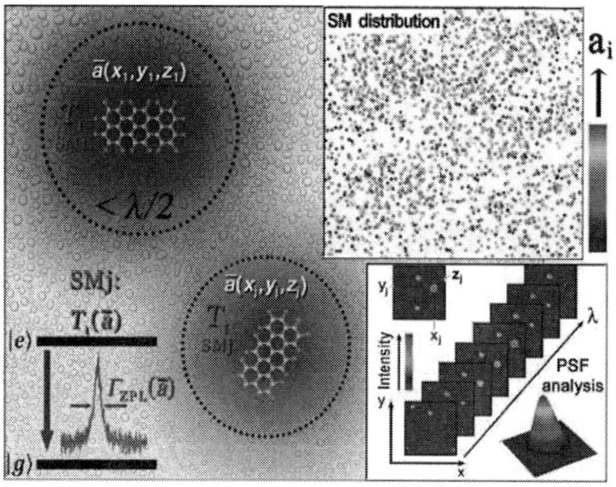

Fig. 1. The sketch to the idea of "multicolor" far-field ZPL-based fluorescence nanoscopy.

This talk discusses the main techniques for detection of single point emitters (chromophore organic molecules, semiconductor quantum dots), temporal and spectral characteristics of SMs and QDs in different environments, the blinking phenomenon, quenching and enhancement of the luminescence of single emitters. There is a discussion of the nanoscopy methods based on the reconstruction of coordinates of single probe emitters, including - 3D-microscopy techniques with the instrumental PSF modification. It is shown how the unique properties of the ZPL can become the basis for a new technique of "multicolor" far-field optical nanodiagnostics of solid media and nanostructures, where the reconstruction of the image is carried out by registering the of the zero-phonon fluorescence images and by restoring the coordinates of a large number of probe SMs with subdiffraction accuracy. In particular, it is proposed to use our approach to nanodiagnostics (mapping) of the local fields in solids, as well as creating a microrefractometry technique [15]. The basis of this approach is the fact that at ultralow temperatures the width of the SM ZPL is determined only by the lifetime of the excited state T_1, which, due to the local field effects, is associated with the effective value of the refractive index of the medium [16, 17]. Thus, the ZPL width value may be converted to an effective value of the refractive index in the local neighborhood of each registered SM.

At room temperature, the idea of "multicolor" nanoscopy can be implemented by methods spectromiroscopy of single colloidal quantum dots. In this case, the coordinates a QD are restored with nanometer accuracy [18], and the additional options for hyperspectral diagnosis can be obtained by analyzing the kinetics and statistics of blinking luminescence of single QDs [19-26]. In particular the idea of microrefractometry with single quantum emitter can be realized with time-resolved spectroscopy of single QD. [27]

Thus, this paper shows that spectromicroscopy of single point emitters can be the basis for the methods of "multicolor" (hyperspectral) nanodiagnostics of condensed matter.

This work was supported by Russian Foundation of Basic Researches (17-02-006525 – development of the microrefractometry with single molecules); Russian Science Foundation (14-12-01415 – application of statistical methods for SMSM researches).

References

1. T. Basché, W.E. Moerner, M. Orrit, U.P. Wild, *Single-molecule optical detection, imaging and spectroscopy* (VCH, Weinheim, Cambridge, 1997)
2. M. Orrit et al., Faraday Discussions **184**, 275 (2015)
3. A.V. Naumov, Physics Uspekhi **56**, 605 (2013)
4. A.V. Naumov, I.Yu. Eremchev, A.A. Gorshelev, Eur. Phys. J. D **68**, 348 (2014)
5. A.M. van Oijen et al., Chem. Phys. Lett. **292**, 183 (1998)
6. S.R.P. Pavani et al., Proc. Natl. Acad. Sci. USA **106**, 2995 (2009)
7. K. K. Rebane, J. Lumin. **100**, 219 (2002)
8. T. Plakhotnik, D. Walser, A. Renn, Phys. Rev. Lett. **77**, 5365 (1996)
9. T. Plakhotnik, E.A. Donley, U.P. Wild, Ann. Rev. Phys. Chem. **48**, 181 (1997)
10. T. Plakhotnik, J. Lumin. **83-84**, 221 (1999)
11. E.A. Donley, T. Plakhotnik, J. Chem. Phys. **114**, 9993 (2001)
12. T. Nonn, T. Plakhotnik, Chem. Phys. Lett. **336**, 97 (2001)
13. A.V. Naumov et al., Angew. Chemie. Int. Ed. **48**, 9747 (2009)
14. A.V. Naumov et al., Phys. Chem. Chem. Phys. **13**, 1734 (2011)
15. T.A. Anikushina et al., Faraday Discussions **184**, 263 (2015)
16. D.V. Kuznetsov, V.K. Roerich, M.G. Gladush, JETP **113**, 647 (2011)
17. D.V. Kuznetsov, V.K. Roerich, M.G. Gladush, Theor. Math. Phys. **168**, 1078 (2011)
18. M.Y. Eremchev, I.Y. Eremchev, A.V. Naumov, EPJ Web of Conf. **103**, 05003 (2015)
19. M. Nirmal et al., Nature **383**, 802 (1996)
20. P. Frantsuzov, M. Kuno, B. Janko, R.A. Marcus, Nature Physics **4**, 519 (2008)
21. A.L. Efros, D.J. Nesbitt, Nature Nanotechnology **11**, 661 (2016)
22. T. Plakhotnik et al., Phys. Rev. Lett. **105**, 167402 (2010)
23. M.J. Fernee et al., J. Phys. Chem. Lett. **3**, 1716 (2012)
24. A.L. Shchukina, I.Yu. Eremchev, A.V. Naumov, Phys. Rev. E **92**, 032102 (2015)
25. I.S. Osad'ko, I.Yu. Eremchev, A.V. Naumov, J. Phys. Chem. C **119**, 22646 (2015)
26. I.Yu. Eremchev, I.S. Osad'ko, A.V. Naumov, J. Phys. Chem. C 120, 22004 (2016)
27. A. Pillonnet et al., Optics Express **20**, 3200 (2012)

EPJ Web of Conferences **132**, 01010 (2017) DOI: 10.1051/epjconf/201713201010
SPECTROSCOPY.SU 2016

Spectroscopy of $RFe_3(BO_3)_4$ multiferroics: phase transitions, spin-phonon interaction, coupled electron-phonon modes

M.N. Popova*

Institute for Spectroscopy RAS, 108840 Troitsk, Moscow, Russia

Abstract. Review of the work performed in the author's laboratory is given, on high-resolution Fourier spectroscopy studies of multiferroics from the family of rare-earth iron borates with the structure of the natural mineral huntite. For these multiferroics, we reveal spectral signatures of interactions between electronic, spin, and lattice degrees of freedom. We have observed and investigated coupled electron-phonon modes in $PrFe_3(BO_3)_4$ and $TbFe_3(BO_3)_4$. The structure of the magnetically ordered phase of $EuFe_3(BO_3)_4$ is determined.

Multiferroics, i.e., materials, in which at least two order parameters coexist, are interesting both for the fundamental solid state physics and in view of possible applications. As a rule, different subsystems (electronic, spin, lattice) of a multiferroic strongly interact with each other. This, in particular, opens a possibility to control an electric polarization by a magnetic field and *vice versa*. The history of multiferroics goes back to the P. Curie's paper of 1894 but only in the middle of the last century Soviet scientists at first predicted in what compounds (named by them "magnetoelectrics") the magnetoelectric effect can be observed and then observed it for the first time. These studies are summarized in the review [1]. The word "multiferroics" appeared in 1994, it allows considering other than magnetic and electric order parameters. At the beginning of this century, a new burst of interest in multiferroics was caused by the synthesis of compounds with a large magnetoelectric effect and by a discovery of new classes of multiferroics (see, e.g., the reviews [2] and [3]). At the same time, linear and nonlinear optical spectroscopy of multiferroics develops; new magneto-optical effects characteristic just for multiferroics are discovered.

In our laboratory, starting from the beginning of 2000-ies, studies of different multiferroics are carried out by the method of the broad-band high-resolution Fourier spectroscopy. Below, we review several our works on the study of new multiferroics from the family of the rare-earth (RE) iron borates with structural type of the natural mineral huntite. These compounds possess a noncentrosymmetric trigonal structure which incorporates helical chains of the FeO_6 octahedra joined by their edges, running along the c axis. The chains are linked by BO_6 triangles and RO_6 prisms. All RE iron borates antiferromagnetically order at $T_N = 30 - 40$ K. Those of them that have an ionic radius of the R^{3+} ion smaller than the one of Sm^{3+} undergo a structural phase transition. We have

* Corresponding author: popova@isan.troitsk.ru

shown that this is a so called "weak" first-order phase transition [4] and have discussed a possible nature of quasi-soft modes [5]. The temperature of the structural phase transition T_s and some of the optical properties turned out to depend on the crystal growth technology [6], which, from one hand, has to be taken into account when interpreting experimental data and, from the other hand, opens a possibility to control T_s.

The antiferromagnetic phase transition is accompanied by noticeable changes in the electronic and phonon spectra of RE iron borates. One observes splittings [7, 8] and shifts [9, 10] of spectral lines corresponding to the $f-f$ optical transitions of RE ions and appearance of forbidden lines [9, 10]. High-resolution electronic spectra of $EuFe_3(BO_3)_4$ allowed us to find that below T_N iron magnetic moments order into a collinear structure along one of the C_2 axes in the *ab* plane [6]. The temperature dependences of the phonon frequencies exhibit peculiarities at T_N, which testifies the spin-phonon interaction [4, 5, 11]. Static and dynamic mechanisms of such interaction are discussed and an experimental proof of a role of the static mechanism is given [11].

We have also undertaken a study of the interaction of phonons and electronic 4*f* excitations of the RE ions in RE iron borates. A temperature-dependent interference between two types of excitations was observed in $PrFe_3(BO_3)_4$ in which the frequency of 4*f* crystal-field electronic excitation of Pr^{3+} falls into the TO-LO frequency interval of the optical A_2^1 phonon mode near 50 cm^{-1}. Experimental data were explained on the basis of a theoretical model of coupled electron-phonon modes. The fitting procedure revealed the value 14.8 cm^{-1} for the electron-phonon coupling constant. This rather large value points to an essential role played by the electron-phonon interaction in the physics of multiferroics [12]. Subsequent *ab initio* calculations have shown that in the mentioned A_2^1 mode translations of the RE ion dominate [13]. Considerably more complicated picture was observed in the spectra of $TbFe_3(BO_3)_4$ in which doubly degenerate lattice phonon and 4*f* electronic crystal-field excitation of Tb^{3+} fall into resonance. An electron-phonon interaction causes delocalization of the CF excitation, an observable Davydov splitting, and formation of coupled electron-phonon modes [5].

The author thanks her colleagues who participated in this research. Financial support by the Russian Science Foundation under Grant No. 14-12-01033 is acknowledged.

References

1. G.A. Smolenskii, I.E. Chupis, Soviet Physics-Uspekhi **25**, 475 (1982)
2. S.-W. Cheong, M. Mostovoy, Nature Mater. **6**, 13 (2007)
3. A.P. Pyatakov, A.K. Zvezdin, Physics-Uspekhi **55**, 557 (2012)
4. D. Fausti et al., Phys. Rev. B **74**, 024403 (2006)
5. S.A. Klimin et. al., Phys. Rev. B **93**, 054304 (2016)
6. M.N. Popova et al., Phys. Rev. B (to be published)
7. M.N. Popova et al., Phys. Rev. B **75**, 224435 (2007)
8. M.N. Popova et al., JETP **118**, 111 (2014)
9. M.N. Popova et al., Phys. Rev. Lett. **102**, 187403 (2009)
10. M.N. Popova et al., Phys. Rev. B. **80**, 195101 (2009)
11. K.N. Boldyrev et al., Phys. Lett. A **376**, 2562 (2012)
12. K.N. Boldyrev et al., Phys. Rev. B. Rapid Comm. **90**, 121101(R) (2014)
13. V.A. Chernyshov et al., Physics of the Solid State **58**, 1642 (2016)

Spatial ordering and abnormal optical activity of DNA liquid-crystalline dispersion particles

S.V. Semenov[2,*], and Yu.M. Yevdokimov[1]

[1]Engelhardt Institute of Molecular Biology RAS, 119991 Moscow, Russia
[2]National Research Center "Kurchatov Institute", 123182 Moscow, Russia

Abstract. In our work, we investigate physicochemical and optical properties of double-strand DNA dispersions. The study of these properties is of biological interest, because it allows one to describe the characteristics of certain classes of chromosomes and DNA containing viruses. The package pattern of DNA molecules in the dispersions particles (DP) is examined. The consideration of the DNA liquid-crystalline DP optical activity based on the theory of electromagnetic wave absorption by large molecular aggregates has been performed. The investigation is also focused on various effects induced by the interaction between biological active compounds and DNA in the content of liquid-crystalline DP.

The nature of ds DNA dispersions properties is interesting from at least two points of view. From the theoretical point of view, the interest is caused by the fact that the physicochemical properties of dispersion particles with size 100 - 1,000 Å can significantly differ from typical continuous solid phases' properties. The difference in the properties evolves as a result of "size effect". Such an effect is caused by the free energy attributed to the surface tension of particles and the existence of possible defects in molecule packing in the particles. The physicochemical properties of liquid-crystalline dispersions of nucleic acids are of biological interest because such particles open a gateway to describing the properties of *Protozoan* chromosomes and DNA-containing viruses, which are isolated microscopic systems with an ordered but labile packing. It is well known that the phase exclusion of the low molecular mass, rigid, linear, ds DNA (molecular mass $<1\times10^6$ Da) is realized under intensive mixing (stirring) of equal volumes the water-salt solutions containing poly(ethyleneglycol), PEG, with water-salt solutions containing a very low (5-10 μg ml^{-1}) concentration of ds DNA at room temperature. As a result the formation of the liquid-crystalline dispersions (LCDs) of these molecules takes place [1]. The particles of the low molecular mass ds DNA dispersions are microscopic droplets of concentrated DNA solution, which cannot be taken in hand (The mode of the ds DNA packing in the LCD particles can model the packing of ds DNA in the primitive biological objects above). Considering the properties of the LCD particles we should keep in mind the following facts: 1) The polymer (PEG) is not included in the content of the formed particles. 2) The size (diameter) of the ds DNA LCD particle is determined by a fine balance between the free energy of these particles and their surface free energy. 3) The packing density of ds DNA molecules in the particles of a dispersion prepared by of mixing of DNA and PEG solutions is determined by the lateral interaction between neighboring DNA molecules at

* Corresponding author: Semenov_SV@nrcki.ru

the moment of their close approaching. The distance between ds DNA molecules is determined by a balance between the repulsive intermolecular forces, and the compressing osmotic pressure of the PEG solution. The minimization of the excluded volume of neighboring linear, rigid, ds DNA molecules induces the parallel, unidirectional (nematic-like) alignment of ds DNA molecules. 4) The maximal density of ds DNA molecules in the LCD particles can be achieved under hexagonal packing. However, this is not a true crystal, because there are disordered water molecules between the ds DNA molecules. The ds DNA molecules possess some disorder around their positions; they can slide and bent with respect of each other, as well as they can rotate around their long axis. It corresponds to the liquid character of their packing. In hexagonal structure (as well as in a case of a hexagonal phase which is formed as a result of an increase in the DNA concentration in a water-salt solutions), we admit the existence of three kinds of orientationally ordered molecular layers. Double-stranded DNA molecules are oriented in a plane of these layers, whose thickness is on the order to the inter-molecular spacing. 5) The ds DNA molecules possess of several levels of chirality (helical secondary structure of ds DNA molecules, helical distribution of counter-ions near DNA surface as well as asymmetry of C-atoms in the sugar residues). Therefore, chiral and anisotropic properties of ds DNA molecules favor to helical twisting of the adjacent DNA molecules packed into the particles of the dispersion. Hence, unidirectional alignment of the ds DNA molecules compete with the tendency of these molecules to form spatially twisted structure of the LCDs. Here one can remind old statement made by Vainshtein and Chistyakov [2], that "we cannot describe directly the spatial distortions of molecules in three dimensions (only by X-ray scattering for liquid crystals) and indirect methods can be used to take into account rotational freedom of molecules about their long axes". To describe the ds DNA molecules spatial packing in LCD we used the small-angle X-ray scattering [3] and also the circular dichroism (CD) [4].

Theoretical calculations of CD of the ds DNA LCD particles is reasonable to perform by means of the approach based on the consideration of electromagnetic wave absorption by large molecular aggregates (MA) [4, 5]. The optical properties of ds DNA LCD, considered as MA, are determined by long-distance interactions between chromophores in a given system. To study the optical characteristics of the structure, a ds DNA LCD particle can be divided into cells and each cell can be considered as a single chromophore. These efficient chromophores, composed by the DNA nitrogen base pairs absorbing dipoles, have an ordering similar to a cholesteric helix with the pitch P. The electric field at the point where a particular chromophore occurs is a superposition of the field of incident light and the electric fields generated by induced dipoles of all other chromophores of the system. As a result, light absorption is a process that involves the total molecular aggregate, and the electric field within the particle is obtained by solving a self-consistent system of the linear equations relative to the field amplitudes at each efficient chromophore. The full extinction cross section, necessary for the CD computations, is given by the optical theorem. To determine vector scattering amplitude the Maxwell's equations for the system under examination are directly solved, applying free-space tensor Green's function. As a result the obtained expression for extinction cross section gives the possibility to calculate the CD of ds DNA LCD as a function of different physical parameters, such as pitch P, particle size D, etc., as well as the change in ordering due to DNA interaction with golden nanoparticles [4,5]. It allows one to describe the peculiarities of the CD of ds DNA LCD complexes with different antibiotics, which is essential for creation of biosensors.

References

1. Yu.M. Yevdokimov, V.I. Salyanov, S.V. Semenov, S.G. Skuridin, *DNA Liquid-Crystalline Dispersions and Nanoconstructions* (CRC Press, New York, 2011)
2. L.V. Azároff, Mol. Cryst. Liq. Cryst. **60**, 73 (1980)
3. E.V. Shtykova et al., Eur. Biohys. J. **39**, 1313 (2010)
4. S.V. Semenov, Yu.M. Yevdokimov, Biophysics **60**, 188 (2015)
5. D. Keller, C. Bustamante, J. Chem. Phys. **84**, 2961 (1986)

EPJ Web of Conferences **132**, 01012 (2017) DOI: 10.1051/epjconf/201713201012
SPECTROSCOPY.SU 2016

Coherent population trapping: its physics and historical roots

A.M. Shalagin*

Institute of Automation and Electrometry SB RAS, 630090 Novosibirsk, Russia

Abstract. There presented basic information on coherent population trapping (CPT) and related phenomenon of electromagnetically induced transparence, 'dark' and 'bright' (anomalous absorption) states is presented. Spectrum transformation is considered in probe field spectroscopy for Λ-scheme in various settings. The close relation is revealed between the effect of CPT in the manifold of magnetic sublevels and the phenomenon of optical pumping (optical orientation and optical alignment). CPT is shown to be generalized to the case of several transitions with a common upper level and with the corresponding number of light fields each being in quasiresonance with its 'own' transition.

When discussing the phenomenon known as coherent population trapping (CPT) or electromagnetically induced transparency (EIT), they usually cite [1] as the first experimental observation and [2] as the first review of theory and experiments from 1976 up to that date. CPT is still in focus due to its expected applications in fine optical frequency standards, ultra-sensitive magnetometry, high-resolution spectroscopy etc.

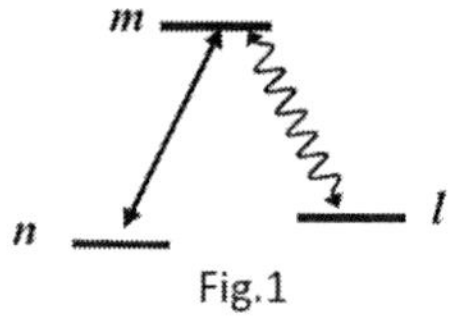

Fig.1

The essence of CPT is as follows. Let us consider three-level Λ-scheme (see Fig.1) where the lower levels l and n do not relax and both the transitions $m - n$, $m - l$ are driven by two quasi-resonant monochromatic fields. In the most simple case, one of the fields (pump field) has arbitrary intensity and the second is weak (probe field). The absorption spectrum of the probe field

$$I(\varepsilon) = I_0 \frac{\Gamma_{ml}^2}{\Gamma_{ml}^2 + (\Omega + \varepsilon - |G|^2/\varepsilon)^2}; \quad \varepsilon \equiv \Omega_\mu - \Omega.$$

Here Ω and Ω_μ – the frequency detunings of the pump and probe fields with regard to the respective transitions, Γ_{ml} – the $m - l$ polarization decay rate. It is evident that under the condition of equal detunings (ε=0) the probe field's absorption is absent. That one of the pump field is absent as well due to the total population being at l. This is the CPT phenomenon: the disappearance of absorption of fields which drive the adjacent transitions. Examples of absorption spectrum $I(\varepsilon)$ are depicted in Fig. 2. CPT yields narrow resonances. When $\Omega = 0$ (Fig. 2*a*) the resonance appears as a narrow gap in center of a broad line (with

* Corresponding author: shalagin@iae.nsk.su

FWHM Γ_{ml}), for large Ω (Fig. 2*c*) it is still a more narrow peak slightly shifted from $\Omega_m = \Omega$. This peak can be used in high-resolution spectroscopy.

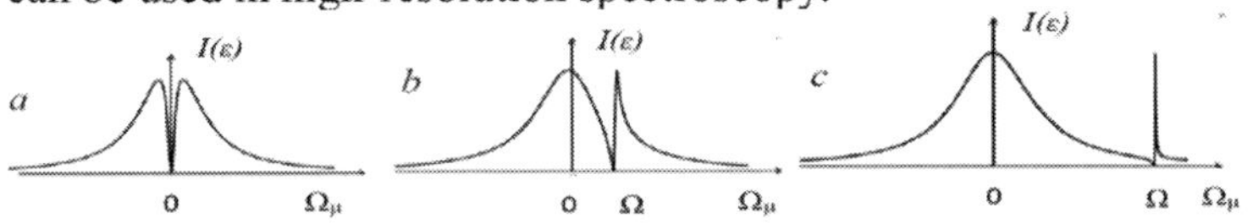

Fig. 2

The condition $\Omega_\mu = \Omega$ provides zero absorptions for arbitrary relation between the fields' intensities. The level *m* is then empty whereas the populations and their coherence

$$\rho_{nn} = \frac{|G_\mu|^2}{|G|^2 + |G_\mu|^2}N; \quad \rho_{ll} = \frac{|G|^2}{|G|^2 + |G_\mu|^2}N; \quad \rho_{nl} = \frac{-G^* G_\mu}{|G|^2 + |G_\mu|^2}N; \quad |\rho_{nl}| = \sqrt{\rho_{nn}\rho_{ll}}\,.$$

The populations depend on the fields' intensities (more precisely, on the Rabi frequencies). The inter-level coherence is maximal. This gives rise to the name of CPT.

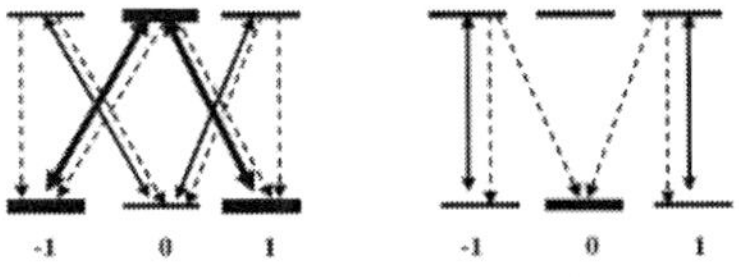

Fig.3

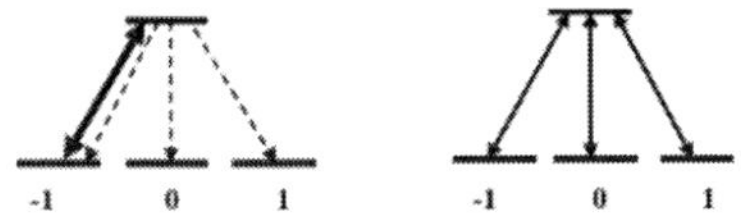

Fig.4

Now we consider the case of polarized radiation and levels with magnetic degeneracy. The transitions between triply degenerate levels (with unit angular momentum) due to linearly polarized radiation are shown in Fig. 3. When the quantization axis is directed along the radiation wave-vector the transitions are those in Fig. 3a. The solid arrows show induced transitions, the dotted arrows stand for spontaneous transitions. Note that transitions between sublevels with zero projection of angular momentum are strictly forbidden. The induced and spontaneous transitions completely depopulate the central sublevel of the ground state as well as the side sublevels of the excited state. As a result, only 'super-solid' transitions and sublevels will contribute to interaction. That means that the standard Λ-scheme of CPT is extracted. The fields driving the adjacent transitions have equal detunings. Hence, there must be zero absorption. This assertion gets almost straightforward if the quantization axis is directed along the electric field. In this case the scheme of transitions is depicted in Fig. 3*b*. The induced and spontaneous transitions cause the optical pumping of the central ground sublevel, so that the absorptions disappears. This example shows CPT and optical pumping being the same in their origin. On the other hand, optical pumping had been already known before the laser invention. It brought Nobel Prize to A. Kastler. We ought to give CPT researchers credit for they pointed a proper choice of basis clarifies the phenomenon as the formation of a specific superposition state which decuples the system from interaction with the fields (the so-called 'dark' state).

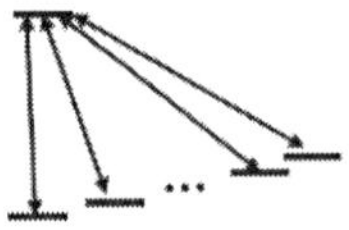

Fig.5

Let us examine another interesting situation with degenerate levels when transitions from triply degenerate ground level to non-degenerate excited one is induced by circularly polarized radiation (see Fig. 4). With the quantization axis along the wave-vector the optical transitions will eventually depopulate the left ground sublevel (see Fig. 4*a*). The central and right sublevels get populated (optical orientation [3]), and the field's absorption disappears. The same phenomenon, being considered with an orthogonal quantization axis, looks like an interaction of the system with three fields with the same frequency (see Fig. 4*b*). A more general situation is imaginable with any number of transitions from a common upper level and with the corresponding number of driving fields (Fig. 5). If at least two fields get equal detunings, the medium will be transparent for all the fields' manifold.

References

1. G. Alzetta et al., Nuovo Cimento Soc. Ital. Fis. B **36**, 5 (1976)
2. E. Arimondo, Prog. Opt. **35**, 257 (1996)
3. A. Kastler, *Preprint les Prix Nobel en 1966* (Nobel Foundation, Stockholm, 1967)

EPJ Web of Conferences 132, 02001 (2017) DOI: 10.1051/epjconf/201713202001
SPECTROSCOPY.SU 2016

High resolution THz gas spectrometer based on semiconductor and superconductor devices

V. Anfertev[1,*], V. Vaks[1], L. Revin[1], I. Pentin[2], I. Tretyakov[1,2], and G. Goltsman[1,2]

[1]Institute for Physics of Microstructures RAS, 603950 Nizhny Novgorod, Russia
[2]Moscow State Pedagogical University, 119991 Moscow, Russia

Abstract. The high resolution THz gas spectrometer consists of a synthesizer based on Gunn generator with a semiconductor superlattice frequency multiplier as a radiation source, and an NbN hot electron bolometer in a direct detection mode as a THz radiation receiver was presented. The possibility of application of a quantum cascade laser as a local oscillator for a heterodyne receiver which is based on an NbN hot electron bolometer mixer is shown. The ways for further developing of the THz spectroscopy were outlined.

The high resolution THz gas spectroscopy method is of great importance for different analytical applications, because the range contains the strongest rotation gas absorption lines. Practical realization leads to strict requirements for THz radiation sources and receivers. The radiation source must be of high frequency stability (relative instability of 10^{-8}-10^{-9}) and widely tunable [1]. The radiation source consists of a frequency synthesizer based on Gunn generator (112-115 GHz) and a semiconductor superlattice (SSL) frequency multiplier. The SSL diode that is used in the frequency multiplier was grown with the molecular beam epitaxy method on GaAs/AlAs materials and is the sequence of 18 periods of alternating layers of GaAs and AlAs [2, 3]. As a consequence of that the high number (up to 17) harmonics generated by the SSL have quite little power (less than nW), a low noise and high sensitive THz receiver is necessary. We applied the NbN superconductive hot electron bolometer (HEB) operating in the direct-detection mode [4, 5].

The test measurements of the water and deuterium water vapor spectral absorption lines with the spectrometer were performed at 1456–1495 GHz, 1680–1725 GHz and 1904–1955 GHz frequency bands. One of the absorption line spectra is presented at Fig. 1.

A THz quantum cascade laser (QCL) [6] is prospective source of coherent radiation in the THz range. Phase locking (PLL) is needed for practical implementation of QCL in high resolution spectroscopy. We used 2 THz continuous wave (CW) mode QCL with output power around 1 mW. The carried out testing measurements showed the possibility of QCL phase locking and implementing for elaboration of a THz gas spectrometer. The QCL structure was soldered to copper plate holder, which was attached to a cold plate of a closed cycle cryostat. The QCL was operated at CW mode at the holder temperature about 15 K.

The QCL radiation was collimated with external optics and then its radiation power was reflected by abeam splitter (3% reflectance) into a THz transparent optical window of a

---

* Corresponding author: anfertev@ipmras.ru

 DOI: 10.1051/epjconf/201713202001

bath helium cryostat to pump a HEB mixer. The QCL worked as a local oscillator for the HEB mixer.

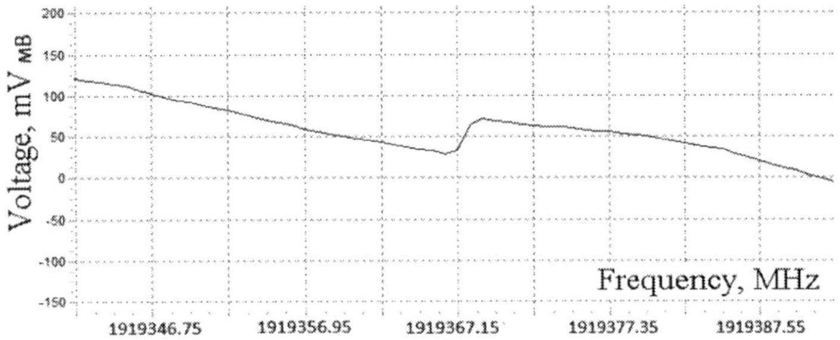

Fig. 1. Water vapor absorption line at 1.919 THz (the 17th harmonic of the synthesizer).

Another radiation was generated with an SSL based multiplier pumped with a backward wave oscillator (BWO) synthesizer, which had frequency stability about 10^{-8} that was much better than the QCL instability. It was transmitted directly through the beam splitter and was used as a probe signal for the HEB mixer.

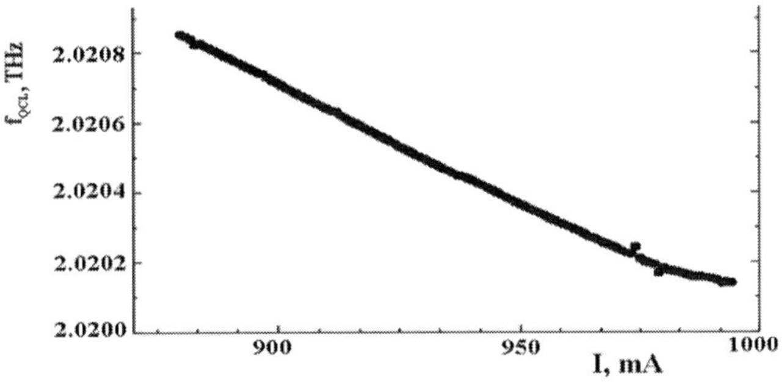

Fig. 2. QCL frequency tuning versus supply current.

The 13th harmonic of the BWO synthesizer had frequency about the same as of QCL. The intermediate frequency signal that resulted from beats of the 13th SSL harmonics and the QCL radiation was studied. We found that the dependence of the QCL frequency upon supply current is monotone and linear (Fig. 2), without instantaneous mode switching in our temperature range and therefore our 2 THz QCL sample was suitable for phase locking [7].

We demonstrated a high resolution THz gas spectrometer for gas studying. We also showed that the QCL is prospective for THz spectrometer developing.

This work was financially supported by RSF (Grant No 15-12-10035).

References

1. V. Vaks, J. of IR, MM and THz Waves **33**, 43 (2012)
2. V.L. Vaks et al., Radiophysics and Quantum Electronics **52**, 511 (2009)
3. D.G. Paveliev et al., Semiconductors **46**, 121 (2012)
4. S. Seliverstov et al., IEEE Transact. on Appl. Superconductivity **25**, 2300304 (2015)
5. I. Tretyakov et al., Appl. Phys. Lett., **98**, 033507 (2011)
6. R. Köhler, Nature **417**, 156 (2002)
7. D.J. Hayton et al., Appl. Phys. Lett. **103**, 051115 (2013)

EPJ Web of Conferences **132**, 02002 (2017) DOI: 10.1051/epjconf/201713202002
SPECTROSCOPY.SU 2016

Single photon transport by a moving atom through sub-wavelength hole

A.E. Afanasiev[1,*], P.N. Melentiev[1], A.A. Kuzin[1,2], A.Yu. Kalatskiy[1,2], and V.I. Balykin[1]

[1]Institute for Spectroscopy RAS, 108840 Troitsk, Moscow, Russia
[2]Moscow Institute of Physics and Technology, 141700 Dolgoprudny, Russia

Abstract. The results of investigation of photon transport through the subwavelength hole in the opaque screen by using single neutral atom are represented. The basis of the proposed and implemented method is the absorption of a photon by a neutral atom immediately before the subwavelength aperture, traveling of the atoms through the hole and emission of a photon on the other side of the screen. Realized method is the alternative approach to existing for photon transport through a subwavelength aperture: 1) self-sustained transmittance of a photon through the aperture according to the Bethe's model; 2) extra ordinary transmission because of surface-plasmon excitation.

The simplest way to transport of photon energy from the one side of opaque screen to another is using of photon transmission through the hole in the screen. Unfortunately as was shown by Bethe [1] and Bouwkamp [2] the efficiency of such transport is low in the case of subwavelength holes. A remarkable discovery about the transmission of light through a subwavelength hole in a metal screen of a finite thickness and a finite conductivity that has been made by Ebbesen et al. [3] has shown that the standard theory of diffraction by small holes is invalid, and, in this case, the transmission of light through the subwavelength hole can be strongly enhanced. The majority of researchers agree that the central role in this phenomenon is played by surface waves, such as surface plasmons.

We have proposed and investigated [4] a fundamentally different mechanism by which a photon can be transferred through a subwavelength hole. It is based on the photon transport that involves the participation of a particle other than a plasmon, namely, a neutral atom. In this scheme, a single atom transfers a single photon through a nanohole. Besides the using of a new particle for photon transport it opens up a new possibilities for surface science. It is possible to use such scheme for investigation of van der Waals interaction of atom inside the hollow cylinder [5]. Another application of the described scheme it could be useful for atom – plasmon interaction investigation. Indeed the subwavelength hole is a highly nonlinear plasmonic element. So the interaction of excited atom with such structure opens new way for tailoring the spectral properties of materials [6].

The basic idea of the single photon transport by a moving atom is presented in Fig. 1. An atom moving toward a metal screen with a hole absorbs a photon of laser radiation immediately in front of the hole. If the lifetime of the atom is substantially larger than the

* Corresponding author: afanasiev@isan.troitsk.ru

time of flight of the atom through the subwavelength hole, the transition of the atom from the excited state to the ground state with emission of a photon can occur under certain conditions on the other side of the screen, which means the transfer of the energy of the photon through the subwavelength hole.

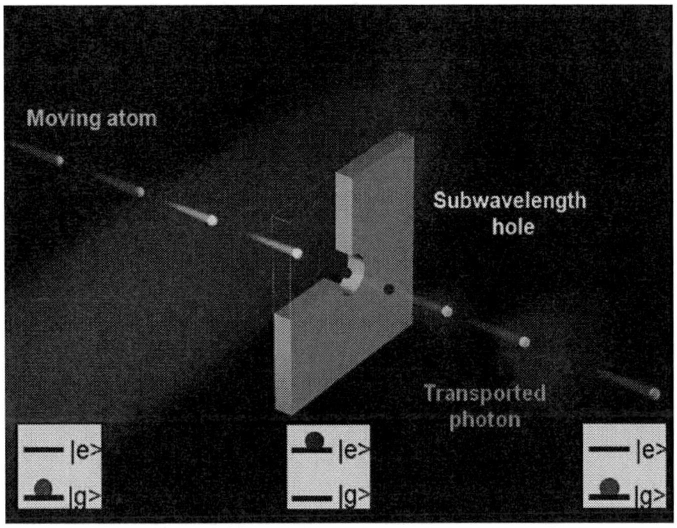

Fig. 1. Basic idea of single photon transport by a moving atom. An atom that moves toward a screen with a nanohole is excited in front of the screen. The long lifetime of the excited atom enables this atom to transfer the excitation energy through the nanohole and reemit it as a photon on the other side of the screen.

The photon transfer efficiency depends on the geometrical dimensions of the hole, the material of the screen with holes, and the velocity and the scheme of energy levels of the atom. At small sizes of the hole, the photon transfer efficiency decreases substantially because of the interaction of the excited atom with the surface, as a result of which the surface of the nanohole in the screen causes deexcitation of the atomic state.

This work was supported by the Program of the Presidium of the Russian Academy of Sciences.

References

1. H.A. Bethe, Phys.Rev. **66**, 163 (1944)
2. C.J. Bouwkamp, Rep. Prog. Phys. **17**, 35 (1954)
3. T.W. Ebbesen, H.J. Lezec, H.F. Ghaemi, T. Thio, P.A. Wolff, Nature **391**, 667 (1998)
4. A.E. Afanasiev, P.N. Melentiev, A.A. Kuzin, A.Yu. Kalatskiy, V.I. Balykin, New J. Phys. **18**, 053015 (2016)
5. A. Afanasiev, V. Minogin, Phys. Rev. A **82**, 052903 (2010)
6. L. Stern, M. Grajower, U. Levy, Nat. Commun. **5**, 4865 (2014)

Theory of propagation of spectrum and correlations of radiation in optically dense gas in the case of the closed excitation contour

K.A. Barantsev*, A.N. Litvinov, and E.N. Popov

Peter the Great Saint-Petersburg Polytechnic University, 195251 St. Petersburg, Russia

Abstract. This work is devoted to generalization of the semi-classical theory of interaction of broadband laser radiation with the atomic gas at the room temperature in the cell in the case of the closed excitation contour. The atomic density matrix equations and spectrum and correlations transport equations have been derived for excitation by fluctuating field with Gaussian statistics. It is shown that the spatial oscillations of radiation intensity and atomic density matrix can be excited. It was found that such medium can serve as a filter of incoherent part of the radiation.

In the field of atomic spectroscopy the coherent population trapping (CPT) effect attracts particular attention [1]. This effect arises when a three-level quantum system (atom) is excited by two-frequency laser radiation. When two-photon resonance takes place, the atom switches to not-absorbing "dark" state, and the medium becomes transparent in some spectral range - region of the CPT-resonance. Width of this resonance can be several orders smaller than the natural width of atomic line. This fact allows to apply CPT in such areas as quantum frequency standards [2], magnetometers [3, 4], quantum memory and slow light [5, 6], etc.

Theory which describes interaction of an atomic gas with the two-frequency laser radiation in the condition of CPT-resonance is well designed. So, in the work [7] appearance of the pseudoresonance was considered in the field of two collinear waves. The optical pumping of Cs atoms by the perpendicular laser fields and CPT effect in these atoms were investigated theoretically and experimentally in [8]. The theory of Dicke narrowing for the CPT-resonance in the cell with buffer gas was built in [9].

The other series of investigations was devoted to CPT effect which is controlled by microwave field forming the closed excitation contour. It was shown that in this case it is possible both destroy and restore CPT by changing in the total phase of the fields [10, 11]. In [12] possibility to control absorption and refraction of the medium by changing in the total phase of the control radiation was demonstrated.

The works [10-12] use the atomic density matrix equations in the approximation when each frequency component of radiation is monochromatic. But in reality, the radiation always has the amplitude and frequency fluctuations. It leads to broadening of the

* Corresponding author: kostmann@yandex.ru

frequency components. In the present work the generalization of the theory of interaction of multifrequency radiation with hot atoms was done in the case of amplitude and phase fluctuations for the closed excitation contour. The equations for transport of energy spectrum and correlations of the radiation were derived for optically dense gas of hot atoms in the cell with buffer gas.

It is demonstrated, that at a certain total phase the spatial oscillations of the radiation intensity and index of refraction appear in the medium. The period and attenuation of these oscillations can be controlled by variation of the microwave field. Propagation of the correlation between the frequency components was analyzed for different degrees of correlation in the input of the medium. It was found that the correlation coefficient can be increased when radiation passes through the medium in the conditions of CPT-resonance. We propose to apply this effect for filtering of incoherent component of the radiation.

This work is supported by State Assignment in science activity for universities (project № 3.1446.2014K, 2014/184), the Russian Foundation for Basic Research (grant № 15-02-01013, 16-32-00587 mol_a), Scholarship of the Government of the Russian Federation, Russian President grant for young candidates of sciences (project MK-6530.2016.2), Federal Special-Purpose Program "Research and development in priority areas of science and technology complex of Russia for 2014-2020 years" (agreement № 14.587.21.0211).

References

1. B.D. Agap'ev, M.B. Gornyi, B.G. Matisov, Yu.V. Rozhdestvenskii, Physics Uspekhi **36**, 763 (1993)
2. J. Vanier, Appl. Phys. **81**, 421 (2005)
3. M. Stahler, R. Wynands, S. Knappe et al., Opt. Lett. **27**, 1472 (2002)
4. A. Akulshin, A. Celikov, V. Velichansky, Opt. Commun. **84**, 139 (1991)
5. M.D. Lukin, Rev. Mod. Phys. **75**, 457 (2003)
6. A. Imamoglu, J.P. Marangos, Rev. Mod. Phys. **77**, 633 (2005)
7. G. Kazakov, B. Matisov, I. Mazets, G. Mileti, J. Delporte, Phys. Rev. A **72**, 063408 (2005)
8. X. Liu, J. Mérolla, S. Guérandel et al., Phys. Rev. A **87**, 029903 (2013)
9. O. Firstenberg, M. Shuker, A. Ben-Kish et al., Phys. Rev. A **76**, 013818 (2007)
10. D.V. Kosachiov, B.G. Matisov, Yu.V. Rozhdestvensky, J. Phys. B: At. Mol. Opt. Phys **25**, 2473 (1992)
11. S.J. Buckle, S.M. Barnett, P.L. Knight et al., Optica Acta **33**, 1129 (1986)
12. H. Li, V.A. Sautenkov, Yu.V. Rostovtsev et al., Phys. Rev. A **80**, 023820 (2009)

Soft X-ray spectroscopy of transition metal compounds: a theoretical perspective

S.I. Bokarev[1,*], R. Hilal[2], S.G. Aziz[2], and O. Kühn[1]

[1]Institute of Physics, University of Rostock, 8059 Rostock, Germany
[2]Chemistry Department, Faculty of Science, 21589 Jeddah, Saudi Arabia

Abstract. To date, X-ray spectroscopy has become a routine tool that can reveal highly local and element-specific information on the electronic structure of atoms in complex environments. Here, we report on the development of an efficient and versatile theoretical methodology for the treatment of soft X-ray spectra of transition metal compounds based on the multi-configurational self-consistent field electronic structure theory. A special focus is put on the L-edge photon-in/photon-out and photon-in/electron-out processes, i.e. X-ray absorption, resonant inelastic scattering, partial fluorescence yield, and photoelectron spectroscopy, all treated on the same theoretical footing. The investigated systems range from small prototypical coordination compounds and catalysts to aggregates of biomolecules.

Tracing atomic and molecular levels in the course of various physical and chemical processes, X-ray spectroscopy is one of the most powerful tools to access structure and properties of matter in different states of aggregation. X-ray spectroscopic techniques probe the local electronic structure of a particular atom in its environment. This is in contrast to UV/vis spectroscopy, where transitions generally occur between delocalized molecular orbitals. The combination of different absorption, emission (scattering) as well as photo- and autoionization X-ray methods allows addressing various aspects of ultrafast dynamics and identification of short-lived intermediates of catalytic reactions. However, interpretation of complex experimental spectra and verification of experimental hypotheses is a non-trivial task and a powerful first principles theoretical approach that allows for a systematic investigation of a broad class of compounds is needed. An additional complexity stems from strong electron correlation and spin-orbit coupling in the core-excited electronic states.

We aim at the development of an efficient and versatile theoretical methodology for the treatment of soft X-ray spectra of transition metal compounds based on the multi-configurational self-consistent field electronic structure theory combined with a perturbative LS-coupling scheme for spin-orbit coupling [1]. A special focus is put on the L-edge photon-in/photon-out and photon-in/electron-out processes, i.e. X-ray absorption, resonant inelastic scattering, partial fluorescence yield, and photoelectron spectroscopy treated on the same theoretical footing [2,3]. We address the application of the X-ray metal L-edge and ligand K-edge spectroscopy to unraveling electronic structure, nature of chemical bonds, oxidation,

* Corresponding author: sergey.bokarev@uni-rostock.de

and spin-states, the interplay of radiative and non-radiative decay channels in transition metal complexes. The investigated systems range from the small prototypical coordination compounds [4-6] and catalysts [7] to the aggregates of biomolecules [8].

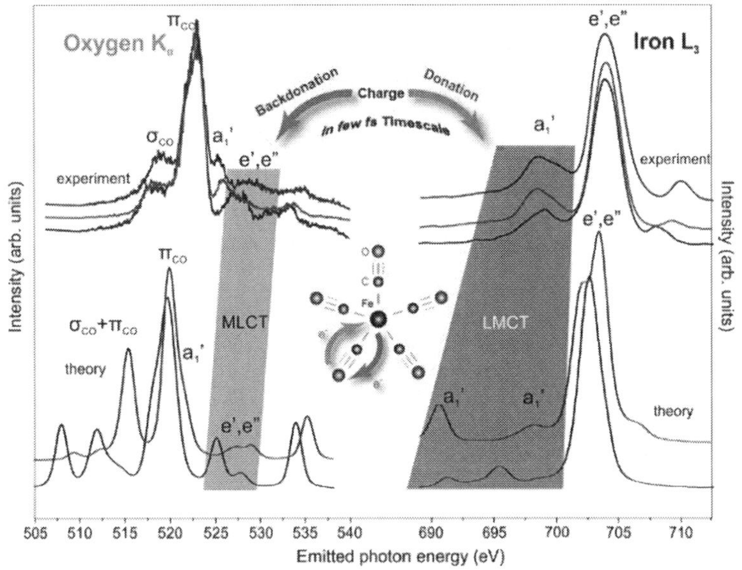

Fig. 1. Experimental and theoretical resonant inelastic X-ray scattering spectra of $Fe(CO)_5$ in solution [4]: Fe L_3-edge (right) and O K-edge (left). The charge-transfer peaks (shaded) provide information on the strength of covalent metal–ligand chemical bonds in terms of donation / backdonation mechanism.

The work is conducted in close collaboration with the experimental group of E.F. Aziz (HZB, Berlin). We acknowledge financial support by the Deanship of Scientific Research (DSR), King Abdulaziz University, Jeddah, (grant No. D-003-435).

References

1. S.I. Bokarev, M. Dantz, E. Suljoti, O. Kühn, E.F. Aziz, Phys. Rev. Lett., **111**, 083002 (2013)
2. G. Grell, S.I. Bokarev, B. Winter, R. Seidel, E.F. Aziz, S.G. Aziz, O. Kühn, J. Chem. Phys. **143**, 074104 (2015)
3. R. Golnak, S.I. Bokarev, R. Seidel, J. Xiao, G. Grell, K. Atak, I. Unger, S. Thürmer, S.G. Aziz, O. Kühn, B. Winter, E.F. Aziz, Sci. Rep. **6**, 24659 (2016)
4. E. Suljoti, R. Garcia-Diez, S.I. Bokarev, K.M. Lange, R. Schoch, B. Dierker, M. Dantz, K. Yamamoto, N. Engel, K. Atak, O. Kühn, M. Bauer, J.-E. Rubensson, E.F. Aziz, Angew. Chem. Int. Ed. **52**, 9841 (2013)
5. N. Engel, S.I. Bokarev, E. Suljoti, R. Garcia-Diez, K.M. Lange, K. Atak, R. Golnak, A. Kothe, M. Dantz, O. Kühn, E.F. Aziz, J. Phys. Chem. B **118**, 1555 (2014)
6. K. Atak, S.I. Bokarev, M. Gotz, R. Golnak, K.M. Lange, N. Engel, M. Dantz, E. Suljoti, O. Kühn, E.F. Aziz J. Phys. Chem. B **117**, 12613 (2013)
7. S.I. Bokarev, M. Khan, M.K. Abdel-Latif, J. Xiao, R. Hilal, S.G. Aziz, E.F. Aziz, O. Kühn, J. Phys. Chem. C **119**, 19192 (2015)
8. M. Preuße, S.I. Bokarev, S.G. Aziz, O. Kühn, Struct. Dynamics **3**, 062601 (2016)

EPJ Web of Conferences **132**, 02005 (2017) DOI: 10.1051/epjconf/201713202005
SPECTROSCOPY.SU 2016

$LiYF_4$:Ho crystal in a magnetic field: high resolution optical spectroscopy

K.N. Boldyrev*, and M.N. Popova

Institute for Spectroscopy RAS, 108840 Troitsk, Moscow, Russia

Abstract. Low-temperature high-resolution optical spectra of $LiYF_4$:Ho in an external magnetic field H are presented. A field-dependent transformation of the hyperfine structure of holmium spectral lines is studied for the cases H$\|c$ and H$\perp c$. We show that Zeeman and hyperfine interactions lead to a strong coupling between crystal-field levels.

The spectra of rare-earth-doped $LiYF_4$ crystals demonstrate uniquely narrow lines, which gives possibility to detect very fine effects. In particular, for $LiYF_4$:Ho crystals, observation and study of the hyperfine structure due to interactions between the 4*f* electrons of the Ho^{3+} ion and nuclear moments of the holmium nucleus [1-3] and isotope effects [4] have been reported. As far as we know, the spectra of $LiYF_4$:Ho subjected to an external magnetic field were not yet studied. In this work, we have performed a high-spectral-resolution optical study of $LiYF_4$:Ho in an external magnetic field. Isotopically enriched $^7Li_{0.1}{}^6Li_{0.9}YF_4$:Ho (0.1 at.%) single-crystal samples were investigated in permanent magnetic fields H$\|c$ and H$\perp c$ (H = 0.6 and 0.9 T).

Figure 1 shows the spectral line corresponding to the allowed electric dipole Γ_{34} ($^5I_8$) $\rightarrow$ $\Gamma_2^{(2)}$ ($^5I_7$) optical transition of the Ho^{3+} ion. Clearly visible additional structure within hyperfine components results from an isotopic disorder within the lithium sublattice [4]. Each of the eight hyperfine components is doubly degenerate in a zero magnetic field [1,2]. A magnetic field splits the Γ_{34} ground state causing a splitting of the spectral line. Interestingly, the intensities of the high-frequency hyperfine components are lower than those of the low-frequency ones, though the Boltzman distribution of populations suggests an inverse situation. The reason for this is that the splitting of the ground state diminishes the energy interval to the first excited state Γ_2 and, thus, enhances an admixture of its wave functions to the wave functions of the ground state. The Γ_{34} ($^5I_8$) $\rightarrow$ $\Gamma_2^{(2)}$ ($^5I_7$) transition borrows its intensity from the strong $\Gamma_2^{(1)}$ ($^5I_8$) $\rightarrow$ $\Gamma_2^{(2)}$ ($^5I_7$) transition, the more the nearer is a given hyperfine sublevel of the ground level to the $\Gamma_2^{(2)}$ ($^5I_7$) level.

An interaction between crystal-field levels, which is enhanced in a magnetic field, manifests itself also in unequal hyperfine intervals in the high- and low-frequency parts of the split line (see Figs. 1 and 2) and in appearance and/or growth in intensity of forbidden lines with $\Delta m = \pm 1$ (here, m is a projection of the nuclear spin).

* Corresponding author: kn.boldyrev@gmail.com

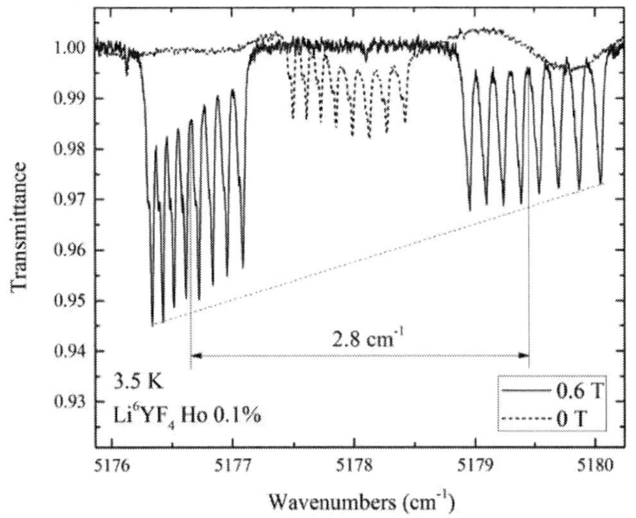

Fig. 1. The π-polarized spectra in the region of the Γ_{34} ($^5I_8$) $\rightarrow$ $\Gamma_2^{(2)}$ ($^5I_7$) transition of LiYF$_4$:Ho^{3+} at 3.5 K in zero magnetic field (dashed trace) and in a magnetic field H$\|c$, $H = 0.6$ T (solid trace). Relative intensities of the hyperfine components simulate a violation of the Boltzman distribution

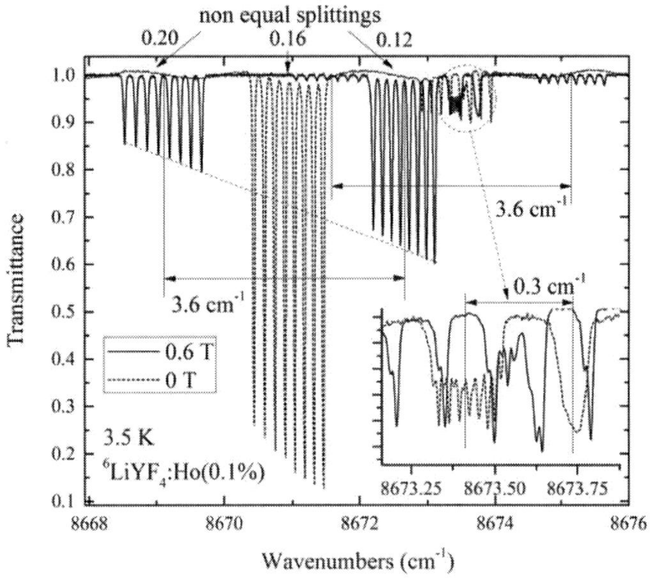

Fig. 2. The π-polarized spectra in the region of the Γ_{34} ($^5I_8$) $\rightarrow$ $\Gamma_2^{(2)}$ ($^5I_6$) transition of LiYF$_4$:Ho^{3+} at 3.5 K in zero magnetic field (dashed trace) and in a magnetic field H$\|c$, $H = 0.6$ T (solid trace), demonstrating a dependence of the hyperfine intervals on the magnetic field. Inset displays a part of the spectrum in the magnetic field $H = 0.9$ T.

This work was performed in the frame of the State Contract № 0039-2014-0005, 8.5.1.

References

1. N.I. Agladze, M.N. Popova, Solid State Comm. **55**, 1097 (1985)
2. N.I. Agladze, E.A. Vinogradov, M.N. Popova, Sov. Phys. JETP **64**, 716 (1986)
3. M.N. Popova, N.I. Agladze, Molecular physics **102**, 1377 (2004)
4. N. Agladze, M. Popova, G. Zhizhin, V. Egorov, M. Petrova, Phys. Rev. Lett. **66**, 477 (1991)

EPJ Web of Conferences **132**, 02006 (2017) DOI: 10.1051/epjconf/201713202006
SPECTROSCOPY.SU 2016

Study of optical and luminescence properties of silicon – semiconducting silicide – silicon multilayer nanostructures

N.G. Galkin, K.N. Galkin*, S.A. Dotsenko, D.L. Goroshko, A.V. Shevlyagin, E.A. Chusovitin, and I.M. Chernev

Institute of Automation and Control Processes FEB RAS, 690041 Vladivostok, Russia

Abstract. By method of in situ differential spectroscopy it was established that at the formation of monolayer Fe, Cr, Ca, Mg silicide and Mg stannide islands on the atomically clean silicon surface an appearance of loss peaks characteristic for these materials in the energy range of 1.1-2.6 eV is observed. An optimization of growth processes permit to grow monolithic double nanoheterostructures (DNHS) with embedded Fe, Cr and Ca nanocrystals, and also polycrystalline DNHS with NC of Mg silicide and Mg stannide and Ca disilicide. By methods of optical spectroscopy and Raman spectroscopy it was shown that embedded NC form intensive peaks in the reflectance spectra at energies up to 2.5 eV and Raman peaks. In DNS with β-FeSi2 NC a photoluminescence and electroluminescence at room temperature were firstly observed.

Semiconductor monolithic and polycrystalline nanoheterostructures created on the base of embedded in silicon matrix the nanocrystals (NCs) of semiconducting silicides (β-$FeSi_2$, $CrSi_2$, h-$CaSi_2$, Mg_2Si), as well as Mg stannide-silicide NCs can have fundamentally new optical properties associated with confinement of electrons and excitons. It is important to develop and create new light emitting and detecting devices and thermoelectrical convertors within silicon complimentary metal – oxide – semiconductor (CMOS) technology. Two- and three-layer structures with embedded β-$FeSi_2$ precipitates (100-150 nm) in silicon with p-n junction were previously grown [1-3], as well as discovered and studied their electroluminescent properties. Decrease of β-$FeSi_2$ NCs sizes down to 40-60 nm and an increase in the number of layers embedded into silicon, have shown promise of this approach in terms of increasing the efficiency of photo- and electroluminescence in grown structures [4,5]. To reduce the probability of mismatch dislocation formation one can use embedding in silicon the nanosize (5-20 nm) crystals of silicides (β-$FeSi_2$, $CrSi_2$, Mg_2Si, h-$CaSi_2$) and Mg stannide, when elastic adjustment of nanocrystal lattice to the silicon lattice or three-dimensional pseudomorphism is proposed. In this case, on the interface semiconductor silicide (stannide) nanocrystals – silicon can expect an absence or significant decrease of dangling bonds and, accordingly, carrier scattering centers and their non-radiative recombination as well as transport of carriers without additional scattering mechanisms.

* Corresponding author: galkin@iacp.dvo.ru

In this paper systematically investigated the nucleation conditions of nanosize silicide's islands of few transition and alkaline earth metals, as well as stannide and magnesium stannide-silicide by *in situ* method of differential reflectance spectroscopy, electron Auger spectroscopy and electron energy loss spectroscopy. It has been established that the formation of nanosize semiconductor islands of semiconductor metal's silicides and stannide leads to peak's formation in the spectra of the change of the imaginary part of dielectric function (CIPDF) in the photon energy range 1.1-1.7 eV [6], which is associated with the formation of the interband transitions with great oscillator force. In case of island formation of metallic silicides a monotonic decrease of CIPDF with increasing photon photons is observed. An optimization of growth parameters has permitted to create monolithic and multilayer nanoheterostructures with embedded in silicon matrix the Fe, Cr and Ca disilicide, Mg and Ca silicide as well as Mg stannide-silicide nanocrystals. By data of cross sectional transmittance electron microscopy it was shown that in monolithic nanoheterostructures (MNHS) an elastic embedding of compressed and extended β-$FeSi_2$ and $CrSi_2$ NCs with 5-20 nm sizes is observed in the monocrystalline silicon without formation of mismatch dislocations [5,7]. In the case of embedding of Mg and Ca silicide and stannide nanocrystals (5-20 nm) in the polycrystalline and amorphous silicon matrix the crystalline structure with arbitrary orientation relatively a silicon matrix is observed. By methods of transmittance and reflectance optical spectroscopy it was established that in the reflectance spectrum the appreciable peaks at energies up to 2.5 eV are formed at increasing of embedded NCs layer quantities [8,9], but in the Raman spectra the peaks correlated with ones of silicides and stannides are observed [10,11]. In MNHS with seven embedded β-$FeSi_2$ NCs and thin silicon interlayers (10-12 nm) the electroluminescence with 25 micro watt power emission in the wavelength range of 1500-1550 nm at 300 K was firstly observed that corresponds to suppression of non-radiative recombination centers, and also to appearance of the quantum Stark effect and formation of excitons with binding energy more than 3/2kT. This intensive electroluminescence was observed at pumping current density of 0.7 A/cm^2, which is a record low value. Linear dependence of optical power emission versus pumping current density throughout the studied range showed the minimization of non-radiative channels for carrier's recombination in diode structures. The stressed structure of β-$FeSi_2$ NCs with sizes lower 15 nm ensured the formation of type-I heterostructure with direct band gap light emission that allows to count on creating of high-speed silicon infrared light emitting diodes working for 300 K.

This work was carried out with financial support from the Russian Science Foundation grant (No. 16-19-10241).

References

1. T. Suemasu, T. Fujii, K. Takakura, F. Hasegawa, Thin Sol. Film. **381**, 209 (2001)
2. T. Suemasu et al., Appl. Phys. Lett. **79**, 1804 (2001)
3. T. Sunokawa et al., Jap. J. Appl. Phys. **44**, 3951 (2005)
4. N.G. Galkin et al., J. Nanosci. & Nanotechn. **8**, 527 (2007)
5. N.G. Galkin et al., Thin Solid Films **519**, 8480 (2011)
6. S.A. Dotsenko et al., e-J. Surf. Sci. & Nanotechnol. **4**, 319 (2006)
7. N.G. Galkin et al., Appl Surf. Sci. **256**, 7331 (2010)
8. N.G. Galkin et al., J. Phys.: Conds. Matt. **19**, 506204 (2007)
9. N.G. Galkin et al., Technic. Phys. **55**, 1036 (2010)
10. S.A. Dotsenko, Phys. Proced. **23**, 41 (2012)
11. N.G. Galkin et al., Jap. J. Appl. Phys. **54**, 07JC06 (2015)

Structure and conformational dynamics of molecules in the excited electronic states: theory and experiment

I.A. Godunov*, V.A. Bataev, D.V. Maslov, and N.N. Yakovlev

Lomonosov Moscow State University, 119991 Moscow, Russia

Abstract. The structure of conformational non-rigid molecules in the excited electronic states are investigated by joint theoretical and experimental methods. The theoretical part of work consist of two stages. In first stage the *ab initio* quantum-chemical calculations are carried out using high level methods. In second stage the vibrational problems of the various dimensions are solved by variational method for vibrations of large amplitude. In experimental part of work the vibronic spectra are investigated: gas-phase absorption and also, fluorescence excitation spectra of jet-cooled molecules. Some examples are considered.

The strategy of the joint theoretical and dynamics of conformational non-rigid molecules in the excited electronic state is presented.

The theoretical part consist of the two stages. In the first stage the *ab initio* quantum chemical calculations are carried out, using high level multiconfigurational methods. The number and types of conformers, adiabatic energies of the excited electronic states, geometrical parameters, vibrational frequencies, values of conformer energy differences and potential barriers of conformational transitions are determined, and in addition — the sections of the potential energy surface (PES) of various dimensions on the coordinates of large amplitude vibrations (torsional, inversional, etc.), which control the conformational dynamics of molecules.

Using these PES sections, in the second stage of theoretical work, the corresponding vibrational problems are solved by variational method and the sets of the energy levels are found.

In experimental part of work the vibronic spectra investigated: gas-phase absorption using multipath optical cell with large length of optical path and also fluorescence excitation spectra of jet-cooled molecules. The energies of conformers in the excited electronic states and vibrational frequencies are determined. In some cases it is possible to estimate the molecular geometrical parameters from the analysis of rotation contours of vibronic bands.

The combination of consistent calculated and experimental data presents quite full picture of the structure and dynamics of conformational non-rigid molecules in the excited electronic states.

* Corresponding author: godunov.msu@gmail.com

The comparison of corresponding theoretical and experimental data demonstrates the potentialities and limits of using computational methods. Nowdays it is the topical and important problem for calculations of structure and properties of molecules in the excited electronic states.

The examples of investigations of propanal (CH_3CH_2CHO) [1-3] and cyclopropanecarboxaldehyde (c-C_3H_5CHO) [4] are considered.

The support from Russian Foundation for Basic Research is acknowledged (Grant No. 16-03-00794a).

References

1. A. Kudich, V. Bataev, A. Abramenkov, V. Pupyshev, I. Godunov, J. of Molecular Structure THEOCHEM **631**, 39 (2003)
2. I. Godunov, N. Yakovlev, A. Abramenkov, V. Bataev, S. Lur'e, Russ. J. Phys. Chem. **79**, 1548 (2005)
3. I.A. Godunov, N.N. Yakovlev, R.V. Terentiev, D.V. Maslov, A.V. Abramenkov, Phys. Chem. Chem. Phys. **18**, 15244 (2016)
4. I.A. Godunov, N.N. Yakovlev, R.V. Terentiev, D.V. Maslov, V.A. Bataev, A.V. Abramenkov, J. Quant. Spectr. Radiat. Trans. **184**, 341 (2016)

Phase transitions and exchange interactions in the $SmCr_3(BO_3)_4$ crystal

E.A. Dobretsova [1,2,*], K.N. Boldyrev[1], M.N. Popova[1], S.Yu. Gavrilkin[2], A.A. Mukhin[3], V.Yu. Ivanov[3], V.V. Mal'tsev[4], N.I. Leonyuk[4], and B.Z. Malkin[5]

[1]Institute for Spectroscopy RAS, 108480 Troitsk, Moscow, Russia
[2]P.N. Lebedev Physical Institute RAS, 119991 Moscow, Russia
[3]A.M. Prokhorov General Physics Institute RAS, 119991 Moscow, Russia
[4]M.V. Lomonosov Moscow State University, 119991 Moscow, Russia
[5]Kazan Federal University, 420008 Kazan, Russia

Abstract. Spectroscopic investigation and specific heat and magnetic susceptibility measurements of $SmCr_3(BO_3)_4$ crystals were performed. The spectra of the Sm^{3+} and Cr^{3+} ions in samarium chromium borate were calculated and parameters of the exchange interactions between the nearest chromium ions, chromium and samarium ions were determined. Three phase transitions were observed at the temperatures T_1= 7.8 ± 0.5 K, T_2 = 6.7 ± 0.5 K, and T_3 = 4.3 ± 0.2 K, their nature is discussed. The crystal structures with different space symmetry groups *R*32 and *C*2/*c* were found to coexist in a $SmCr_3(BO_3)_4$ single crystal.

Samarium chromium borate $SmCr_3(BO_3)_4$ is a member of a large family of rare-earth (RE) borates with structural type of the natural mineral huntite. These compounds are described by a general formula $RM_3(BO_3)_4$, where *R* is Y or La-Lu and *M* is Al, Ga, Fe, Cr, or Sc. The compounds are actively investigated due to their interesting physical properties and a promising application potential. However, RE chromium borates are not so well studied. It is known that $NdCr_3(BO_3)_4$ is an easy-plane antiferromagnet below the temperature 8.0K [1] and that $EuCr_3(BO_3)_4$ undergoes an antiferromagnetic phase transition at $T_N \approx 9$ K [2]. A phase transition at 5 K was supposed to occur in $SmCr_3(BO_3)_4$ [3].

We succeeded to prepare a big oriented single crystal of $SmCr_3(BO_3)_4$ and to perform magnetic measurements on it. It follows from the temperature dependences of the magnetic susceptibility that the magnetic moments of chromium and samarium ions (samarium ions are polarized by the exchange field) are ordered in the *ab*-plane. However, as this sample was not preserved, heat capacity and optical spectra measurements were carried out on other samples. Crystal characterization by X-ray diffraction and infrared (IR) spectroscopy revealed a polytypic nature of $SmCr_3(BO_3)_4$. The crystals can be rhombohedral (space group *R*32) or monoclinic (*C*2/*c*), depending on growth conditions. Moreover, domains of both structural phases can coexist in one single crystal. For measurements we chose the crystals with a predominant content of the *R*32 phase. The temperature dependence of the

* Corresponding author: elenadobretsova89@gmail.com

heat capacity $C(T)$ of $SmCr_3(BO_3)_4$ has revealed two second-order phase transitions, at T_1 = 7.8 K and T_2 = 6.7 K, and the first-order transition at T_3 = 4.3 K.

Optical spectra registered in the range of *f-f* transitions in the Sm^{3+} ions and *d-d* transitions in the Cr^{3+} ions helped to identify energy level positions in these ions in the crystal field (CF) of $SmCr_3(BO_3)_4$. It was also found from the observed spectral lines splitting that the compound orders magnetically at a temperature near 7 K.

The set of experimental data allowed determining CF parameters for the Sm^{3+} and Cr^{3+} ions in $SmCr_3(BO_3)_4$. The *g*-factors, calculated in this paper for the ground orbital singlet of Cr^{3+}, are g_{aa}=1.9648, g_{bb}=1.9675, g_{cc}=1.9702 but the initial CF splitting parameter is D = -0.379 cm^{-1}. When considering magnetic properties of $SmCr_3(BO_3)_4$, we supposed the presence of isotropic exchange interactions between the nearest neighbor chromium and samarium ions ($H_{fd} = -2J_{fd}\boldsymbol{S}_{Cr}\boldsymbol{S}_{Sm}$), chromium ions inside the chains ($H_{12} = -2J_{nn}\boldsymbol{S}_{Cr1}\boldsymbol{S}_{Cr2}$), and the next nearest neighbor chromium ions in the closest chains ($H'_{12} = -2J_{nnn}\boldsymbol{S}_{Cr1}\boldsymbol{S}_{Cr2}$). A projection of the exchange interaction operator of the Sm^{3+} ion with one of the closest Cr^{3+} ions onto the ground state (the lowest doublet) of the samarium ion in the crystal-field leads to a spin Hamiltonian

$$H_{fd}^{(eff)} = -2J_{fd}[G_{\perp}(S_{Cr,x}S_x + S_{Cr,y}S_y) + G_{\|}S_{Cr,z}S_z], \quad (1)$$

where S_α are components of the effective spin of the samarium ion S=1/2, and $G_{\|}$=0.2743, $G_{\perp}$=2.2.

Temperature dependences of the static magnetic susceptibility tensor components and magnetic-field dependences of the magnetization were calculated in the self-consistent field approximation, using a model of clusters (dimers) containing the closest chromium ions in a chain. This model is analogous to the model of dimers in the chains of iron ions, used formerly to investigate magnetic and spectral properties of the RE iron borates [4]. A comparison between the calculated and experimental temperature dependences of the magnetic susceptibility gave the values J_{nn}=-2.8 K and J_{nnn}=0.75 K for the exchange integrals; the value J_{fd}=0.132 K was obtained from the measured samarium ion ground-state splitting in the exchange field at low temperatures on the assumption of a collinear antiferromagnetic structure in the chains of chromium ions.

It follows from the analysis of magnetic characteristics of Sm^{3+} and Cr^{3+} that a hypothesis about a spin-reorientational nature of the low-temperature transition with a rotation of magnetic moments from the *ab* plane to the *c* axis is not valid. The main question about the exchange interaction parameters in comparison with iron borates refers to the exchange interaction between the chromium chains. According to neutron scattering data, the iron ions are ferromagnetically ordered in every *ab* plane, due to the antiferromagnetic interaction (J_{nnn} <0) of the closest chains. Because of the change of sign in the exchange integral J_{nnn} in $SmCr_3(BO_3)_4$, one may expect a magnetic structure essentially different from the one observed in iron borates. Additional investigation is required to determine a magnetic structure of chromium borates.

This work was supported by the Russian Science Foundation (Grant No 14-12-01033). E.A.D. acknowledges a Scholarship of the President of the Russian Federation (SP-754.2015.1).

References

1. E.A. Popova et al., Phys. Rev. B **75**, 054446 (2007)
2. L. Gondek et al., J. Sol. St. Chem. **210**, 30 (2014)
3. K.N. Boldyrev, E.P. Chukalina, N.I. Leonyuk, Phys. Sol. St. **50**, 1681 (2008)
4. M.N. Popova et al., Phys. Rev. B **75**, 224435 (2007)

Detection of fatty product falsifications using a portable near infrared spectrometer

A.V. Kalinin*, and V.N. Krasheninnikov

Institute for Spectroscopy RAS, 108840 Troitsk, Moscow, Russia

Abstract. Spreading sales of counterfeited fatty-oil foods leads to a development of portable and operational analyzer of typical fatty acids (FA) which may be a near infrared (NIR) spectrometer. In this work the calibration models for prediction of named FA were built with the spectra of FT-NIR spectrometer for different absorption bands of the FA. The best parameters were obtained for the wavelength sub-band 1.0-1.8 μ, which includes the 2nd and 3rd overtones of C-H stretching vibrations (near 1.7 and 1.2 μ) and the combination band (1.42 μ). Applicability of the portable spectrometer based on linear NIR array photosensor for the quality analysis of spread, butter and fish oil by the typical FA has been tested.

Determination of individual fatty acids (FA), and the groups of them (cis-, trans-, saturated, unsaturated, and other) in food products and supplements have been made in recent studies [1-2] using FT spectrometer in the near infrared (NIR) spectral region for a variety of overtones and combinations of fundamental vibrations of C-H, C = C, C = O with the use of multivariate calibration (regression) on projections to latent structures (PLS). Efficiency of FA prediction using different absorption bands of overtones and combinations in these works has not been studied, which prevents the selection of optimal technical conditions for creation of portable NIR spectrometer, eg with the linear InGaAs array photosensor. The aim of this work was to study the efficiency of calibration models built with different FA absorption bands, and the applicability of NIR spectrometer with linear G9213S photosensor, Hamamatsu, Japan, as a prototype of a portable quality analyzer of spread, butter and fish oil on content of FA: butyric C4:0, myristic C14:0, palmitic C16:0, oleic C18:1, and the amount of polyene Ω3 C20:5+C22:6.

Spectroscopic analysis involves time-consuming multi-component calibration over a large number of calibration samples. In previous work [3], we have developed cost-effective technique that provides a good calibration model with a special choice of variations of three components in the set of only 35 samples. A similar design was done in the first set of 50 mixtures of milk fat, palm and sunflower oils to determine the content of FA C4:0, C14:0 and C16:0, and in the second set of 112 samples to determine the content of C16: 0, C18: 1 and C20: 5 C22 + 6. Calibration (35, 38 and 39 units) and testing (35 and 25 units, respectively) sets of samples were standardized using the thin-film gas chromatograph Kristallyuks4000M Ltd. Interchrom, RF, in accordance with ISO 5508: 1990 (E) .This samples can serve also as standards of milk fat, palm, olive and fish oil fractions, from which weighed portions they were manufactured. Pretreatment of calibration samples and tested products was performed taking into account the testing

* Corresponding author: kalinin@isan.troitsk.ru

results of several methods for fat extraction, obtained in [2]. Mixtures of oils and fats in the test tube was heated to 60 ° C and held for sedimentation, then the lower fraction was drained through the hole at the bottom, and the top foam, if any, was filtered in a special incubator as described in [3]. FA ranges of variations were in the first set of samples for C4:0 (0.5-4), for C14:0 (6-14), for C16:0 (14-33), in the second set for C16:0 (14- 35), for C18:1 (24-47), and for C22 5+C24:6 (12-25)% of total fat.

Absorption spectra of second set samples were acquired first using FTNIR spectrometer MPA Bruker, Germany in the wavelength range 0.9-2.2 μ, and, secondly, via a fiber-optic InGaAs spectrometer within the range 0.9-1.65 μ (see Fig. 1). Built on two sets of samples PLS models were tested with the samples of not less than 25 standards that were not used in the calibration model. Models with the spectra of the FTNIR spectrometer for four spectral sub-regions indicate the similar errors, but the model for IGA spectrometer in the range 1.1-1.64 μ has the lowest total error, in comparison with the models for full range of FTNIR spectrometer. The best tested models (standard error of 0.49, 1.79 and 0.51 for the C16:0, C18:1 and Ω3, respectively) then were used for analyses of certain foods and supplements, the results are presented in Table 1.

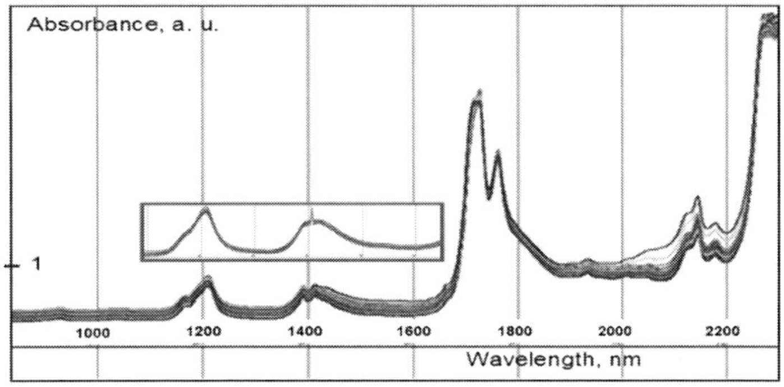

Fig. 1. Absorbance spectra of 51 fat-oil mixtures in the test tube Ø 6 mm obtained by FTNIR spectrometer MPA Bruker. Spectra of the same samples accumulated with IGA spectrometer (insert).

Table 1. Prediction of the content of typical FA in spreads and butter and.Ω3 FA in dietary supplements and fish oils

Value / Product	FA fraction					Conclusion
	C4:0	C14:0	C16:0	C18:1	Ω 3	
Supplement Fish Oil "Ω3"	<.5		2.4	39.9	24.5	Normal
Oil of Raw Capelin	<.5		15.9	41.2	13.1	
Oil of Baked Capelin	<.5		14.2	410.	11.1	Less than 15% raw
Spread «Kremliovskoye»	1.1	6.1	29.2			Shortage of milk fat
Spread «Extra»,72%	<.5	12.2	30.3			Shortage of milk fat
Butter "Valio", 72%	2.2	8.7	32.4			Normal
Butter «Is Vasilcova»	<.5					Falsed
Milkfat substitute "Efko"	1.5	2.6	44.			Milk fat less 30%

Models derived from the spectra accumulated by the IGA spectrometer have the same accuracy and reliability of determining the content of C16:0 and C18:1, compared with the models on the Fourier spectra to estimate the proportion of milk fat in spreads published in [2]. Determination of characteristic FA helped to draw conclusions on the percentage of milk fat in the milk-containing products and the content of FA Ω3 in fish oil.

Conclusions: a) the content of FA C4: 0, C14: 0 and C16: 0 using InGaAs spectrometer were well defined in mixtures where milk fat percentage was of 15% or higher; b) FA Ω3 were well-defined by all models; c) the best parameters for determining FA C16:0, C18:1 using FTNIR spectrometer in the spectral range 0.9-2.2 μ had the model for the sub-band 1.1-1.8 μ, which includes the 2nd and 3rd overtone of C-H stretching vibrations (near 1.7 and 1.2 μ) and combination at 1.42 μ. According to the concepts of atherosclerosis evolution [4] the contents of C16: 0 and C18: 1 FA in food and the serum of patients has special diagnostic significance. Creating a InGaAs spectrometer for the 0.9-1.8μ spectral region, will allow, perhaps, to improve the methodology of FA determination and to realize portable FA analyzer for operational monitoring of fish, meat and dairy products and dishes.

References

1. http://lipidlibrary.aocs.org/topics/IR-trans/index.htm
2. G. Cabassi et al., J. Near Infrared Spectrosc. **21**, 395 (2013)
3. A. Kalinin et al. J. Near Infrared Spectrosc. **21**, 409 (2013)
4. V.N. Titov, Pathogenesis **11**, 18 (2013)

EPJ Web of Conferences **132**, 02010 (2017) DOI: 10.1051/epjconf/201713202010
SPECTROSCOPY.SU 2016

Magnetic phase transitions in two-dimensional frustrated $Cu_3R(SeO_3)_2O_2Cl$. Spectroscopic study

S.A. Klimin[1,*], and I.V. Budkin[1,2]

[1]Institute for Spectroscopy RAS, 108840 Troitsk, Moscow, Russia
[2]Moscow Institute of Physics and Technology, 141700 Dolgoprydny, Russia

Abstract. Using optical study of electronic spectra of rare-earth (RE) ions, magnetic phase transitions in the low-dimensional frustrated RE magnets $Cu_3R(SeO_3)_2O_2Cl$ (R = Sm, Yb, Er, Nd, Pr, Eu) were investigated. Phase transitions were registered either by splittings of crystal-field (CF) doublets or by repulsion of CF levels of f-ions in a staggered magnetic field. Different scenarios of magnetic order in isostructural compounds of the francisite family are discussed.

Magnetic system of a natural mineral francisite $Cu_3Bi(SeO_3)_2O_2Cl$ [1] has attracted attention of scientists due to its unusual geometry. Copper ions in the orthorhombic *Pmmn* crystal structure of francisite form a network of two-dimensional (2D) layers with a buckled Kagome lattice [2]. Due to weak interlayer interactions, an antiferromagnetic (AFM) ordering occurs in francisite at T_N = 24 K [2]. Bismuth can be replaced by a rare-earth (RE) element, while the crystallographic structure remains the same [4]. The family of compounds $Cu_3R(SeO_3)_2O_2Cl$ (R = RE) can be considered as a model system of 2D copper layers with different types of interactions, due to introduction of a second (RE) magnetic subsystem. In the case of R = Y [5] or Sm [6], the Néel temperature is also low (T_N = 35 K), but higher as compared to parent francisite, which is due to a discrepancy in ionic radii of bismuth and RE ions. In addition to AFM ordering, $Cu_3Sm(SeO_3)_2O_2Cl$, experiences a spin-reorientation second-order phase transition at T_R = 9K [6]. In this work, a comparative study of compounds with R = Sm, Yb, Er, Nd, Pr, and Eu is presented. We have carried out infrared spectroscopy study of f-f transitions in RE ions, with the aim to investigate magnetic ordering in compounds with different RE ions.

Transmittance spectra in a wide spectral range (1800 – 12000 cm^{-1}) were measured using a Fourier–spectrometer Bruker IFS125HR. Mixtures of $Cu_3R(SeO_3)_2O_2Cl$ and KBr powders were pressed into tablets, with an effective thickness of the substance ~ 10 mkm. Low temperatures were achieved using a closed-cycle cryostat Cryomech PT403.

The CF energies of six R^{3+} ions were found in the spectral range studied. Spectroscopic data on temperature-dependent low-energy states of RE ions were used to calculate the RE contribution [7, 8] into magnetic susceptibility and heat capacity. Magnetic ordering was detected either by splitting of Kramers (T_N^{Sm}=34 K, Fig. 1 a; T_N^{Yb}=34 K; T_N^{Er}=35 K) or of casual (T_N^{Ho}= 37K, Fig. 1 b; T_N^{Pr}=34 K) doublets or by internal-magnetic-field-induced repulsion of CF levels (T_N^{Eu}=36 K, Fig. 1 c). The temperatures T_N obtained are very close to that in $Cu_3Y(SeO_3)_2O_2Cl$. This fact was expectable, as d-d exchange interactions are the

* Corresponding author: klimin@isan.troitsk.ru

strongest ones in mixed d-f magnetic systems [9]. In the case of RE francisites, the d-d interactions are expected to be approximately the same for all the family, because the geometry of bonds does not change significantly due to closeness of ionic radii of RE ions.

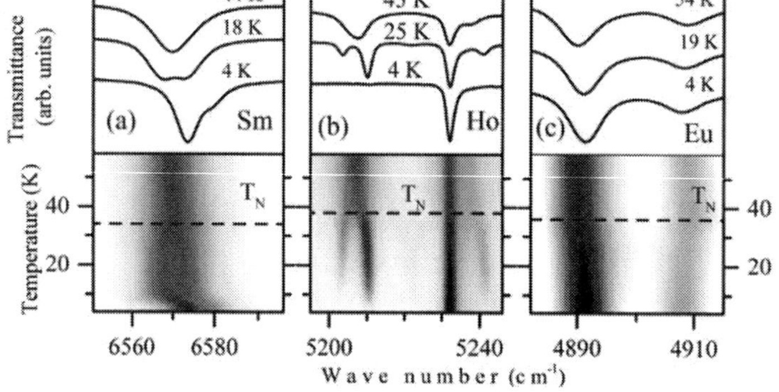

Fig. 1. Transmission spectra of $Cu_3R(SeO_3)_2O_2Cl$, *R*=Sm, Ho, Eu, at several temperatures (upper panels) and intensity maps (lower panels) in the regions of the following multiplets, (a) $^6F_{3/2}$ of Sm^{3+} ion, (b) $^5I_7$ of Ho^{3+} ion, and (c) $^7F_6$ of Eu^{3+} ion.

Francisites with Kramers ions studied demonstrate different temperature behavior of the RE ground doublet (GD). GD of erbium splits at T_N, while GDs of Sm and Yb do not. The last two split only at temperatures $T<T_R$, T_R being 9 K in both cases. To understand the reason of such behavior, we address a relation for the splitting Δ of a doublet state in a staggered magnetic field:

$$\Delta = \sqrt{\sum_{j=x,y,z}(g_j B_{eff,j})^2}, \qquad (1)$$

where *g* is the magnetic *g*-factor of the RE ion. The magnetic structure of copper ions in AFM phase for all the family is expected to be of the same type as in $Cu_3Bi(SeO_3)_2O_2Cl$ [3] and $Cu_3Y(SeO_3)_2O_2Cl$ [4]. It follows from data on the magnetic structure of $Cu_3Bi(SeO_3)_2O_2Cl$ [3] that in the case of symmetrical exchange the effective magnetic field acting on the RE ion is directed along the z-axis of a crystal ($B_{eff}||z$). In this case, Eq. (1) is reduced to $\Delta=g_zB_{eff,z}$. As $\Delta_{GD,Sm(Yb)}=0$ at $T_R<T<T_N$, $g_z=0$ for the ground doublets of both the Sm^{3+} and Yb^{3+} ions. Appearance of a splitting in the temperature range $T<T_R$ means a change of B_{eff}, namely, the appearance of its component along a non-zero component of the magnetic *g*-factor of the RE ion. Thus, we conclude that spin-reorientational transitions take place in $Cu_3Sm(SeO_3)_2O_2Cl$ and $Cu_3Yb(SeO_3)_2O_2Cl$ at T_R. The reason for that is a gain in energy that the system obtains due to the splitting of the ground doublet of the RE ion which exceeds loss in energy due to the reorientation of the spins of copper ions.

The authors are grateful to P.S. Berdonosov and E.S Kuznetsova (Lomonosov Moscow State University), who kindly provided us with the samples.

References

1. A. Pring, B.M. Gatehouse, W.D. Birch, Amer. Mineral. **75**, 1421 (1990)
2. P. Millet, B. Bastide, V. Pashchenko et al., J. Mater. Chem. **11**, 1152 (2001)
3. M. Pregelj, O. Zaharko, A. Günther et al., Phys. Rev. B **86**, 144409 (2012)
4. P.S. Berdonosov, V.A. Dolgikh, Rus. J. Inorg. Chem. **53**, 1353 (2008)
5. K.V. Zakharov, E.A. Zvereva, P.S. Berdonosov et al., Phys. Rev. B **90**, 214417 (2014)
6. K.V. Zakharov, E.A. Zvereva, M.M. Markina et al., Phys. Rev. B **94**, 054401 (2016)
7. E.A. Popova, S.A. Klimin, M.N. Popova et al., J. Magn. Magn. Mater. **331**, 133 (2013)
8. S.A. Klimin, A.S. Galkin, M.N. Popova, J. Alloys Comp. **625**, 193 (2015)
9. M.N. Popova, J. Alloys Comp. **275 –277**, 142 (1998)

Analysis of the parameters of Franck–Condon and Herzberg–Teller interactions the molecules of substituted diphenylbutadienes

V.V. Kompaneez*, and I.A. Vasilieva

Moscow State Pedagogical University, 119991 Moscow, Russia

Abstract. We have done quantitative analysis of parameters Franck-Condon and Herzberg-Teller interactions of substituted diphenylbutadiene with general formula all-trans-X-Ph-(C = C)$_2$-Ph-Y (X, Y = H, NO_2, $N(CH_3)_2$, NH_2 various combinations). The research shows the influence of substituents on the vibronic parameters of characteristic bands, which describe the state (vibrations, types of deformation under excitation) of the phenyl ring and the polyene bridge at the diphenylbutadiene molecule. Results described impact of the substituent's nature on the parameters of intra- and intermolecular interactions presents for the studied compounds.

The methods of site-selective spectroscopy are very powerful for investigations of inter- and intra-molecular processes in complex organic dye doped solids (see e.g. [1-4] and references therein). In particular a lot of information about intramolecular interactions can be obtained by joint analysis of absorption and luminescence electronic-vibrational spectra of organic compounds. [5-8] In the majority works, which devoted to the spectra of substituted polyenes with trans-configuration, for the same substituted compounds there is considerable overlap between the general characteristic vibrational bands locations and these interpretations.

It is known that the substituents in the para position (in this study – nitro and amino groups) irrespective of their electronic nature violate uniform distribution of electron density in the molecule of DPB [9]. It should affect by the quantification of intramolecular interactions. Therefore, it was interest to produce comparative analysis of the spectral properties and quantification of the contribution of intramolecular interactions in the formation of the fine-structure spectra. The relative intensities of bands in the spectra of were calculated by the procedure described in [7, 8]. Based on the modern theory of molecular spectra in the adiabatic approximation ,which makes it possible to simultaneously take into account the influence of the Franck–Condon and Herzberg–Teller interactions on the shape of optical bands [10], we calculated the parameters a and α (of the FC and HT interactions, respectively) for each vibronic transition of the studied molecules.

In this study, the basic oscillations for substituted compounds were correlated with the interpretation of similar value 1,4-diphenylbutadiene, which spectral properties have been studied previously [11-13]. We shown to considered acceptors and donors, and besides the

* Corresponding author: kompaneez@gmail.com

choice of the substituent in the aromatic ring was conditioned by the intention to show the obvious symmetry breaking, as well as variation for symmetry recovery.

The research showed that the parameters of intra- and intermolecular interactions, forming fine-spectra of the selected compounds exhibit a high degree of tolerance in the homologous row of molecules, and the possibility of a fragmented approach by the describing the fundamental bands. The obtained spectra confirm the validity of our findings, in addition the possibility of prediction them via vibronic parameters and select the type of substituent values - for the donor or acceptor interaction per molecule [12, 13].

It has been shown that the implantation of the diphenylpolyene's ring electron-donor or electron-acceptor substituent leads to broadening of the lines and increased background spectra. The implantation of substituents helps reduce the dipole moment of the molecule, which also reduces the value of the intermolecular interaction parameters. The comparative presence donor and acceptor in the different phenolic rings admited to improves symmetry of the conjugate spectra, and significantly increase the internal energy of the molecule. Thus, the implantation of substituents does not affect to the localisation characteristic bands into the spectrum and can be used to strength the weak manifestation of the bands in the spectra of some luminophors.

This work was financially supported by the Russian Foundation for Basic Research, project no. 15-32-21100-mol_a_ved.

References

1. M. Orrit, J. Bernard, R.I. Personov, J. Phys. Chem. **97**, 10256 (1993)
2. A.V. Naumov, Physics Uspekhi **56**, 605 (2013)
3. K.R. Karimullin, A.V. Naumov, J. Lumin. **152**, 15 (2014)
4. A.V. Naumov, I.Y. Eremchev, A.A. Gorshelev, Eur. Phys. J. D **68**, 348 (2014)
5. A.V. Naumov, N.L. Naumova, I.S. Osad'ko, Opt. Spectrosc. **91**, 750 (2001)
6. N.L. Naumova, A.V. Naumov, A.N. Nikitina, I.A. Vasil'eva, Zh.A. Krasnaya, Yu.V. Smirnova, Opt. Spectrosc. **92**, 426 (2002)
7. N.L. Naumova, I.A. Vasil'eva, I.S. Osad'ko, A.V. Naumov, Opt. Spectrosc. **98**, 535 (2005)
8. N.L. Naumova, I.A. Vasil'eva, I.S. Osad'ko, A.V. Naumov, J. Lumin. **111**, 37 (2005)
9. B.M. Krasovitskii, B.M. Bolotin, *Organic Luminophores* (Khimiya, Moscow, 1984)
10. I.S. Osad'ko, *Selective Spectroscopy of Single Molecules* (Springer, Berlin, 2002)
11. I.A. Vasilieva, V.V. Kompaneets, Opt. Spectrosc. **115**, 32 (2013)
12. I.A. Vasilieva, V.V. Kompaneets, Opt. Spectrosc. **114**, 60 (2013)
13. V.V. Kompaneets, I.A. Vasilieva, Bull. RAS: Physics **80**, 848 (2016)

Generation of two-lobe light fields with a rotating intensity distribution under propagation for single emitter spectroscopy

V.G. Volostnikov[1,2], E.N. Vorontsov[1], S.P. Kotova[1,2,*], N.N. Losevsky[1], D.V. Prokopova[1,2], E.V. Razueva[1], and S.A. Samagin[1]

[1]Samara Branch of the Lebedev Physical Institute RAS, 443011 Samara, Russia
[2]Samara National Research University, 443086 Samara, Russia

Abstract. The paper outlines the results of studies on the generation of two-lobe light fields with the intensity distribution rotating during the field propagation. Such fields are needed to determine the depth of bedding of single emitters in spectral studies of substance properties. On the base of the spiral beam optics, the phase distributions were obtained for the synthesis of two-lobe fields with different speeds of rotation of the intensity distribution. The light fields have been formed by using a liquid-crystal spatial phase modulator HOLOEYE HEO-1080P. The influence of the illuminating beam parameters and the aberrations of the system on the quality of the formed light field was also studied.

The problem of accurately determining the position of single fluorescent molecules, quantum dots and other nanostructures arises in the spectral study of physical and chemical properties of substances [1,2]. Meanwhile the point light source coordinates in space are likely to change with time. The task of the exact estimation of the longitudinal coordinate of the emitting nano-size object in microscopy is still unsolved. As a part of this task focuses on the use of light fields on the distribution of the intensity of which can determine the depth of the object. And the two-lobe light fields with their intensity distribution looking like two maxima, are considered to be particularly attractive ones [3].

The idea of the suggested approach for estimating the depth of bedding is as follows. The radiation from the point light emitters passes through the optical system of a microscope and a special phase filter. Essentially this filter is a diffraction optical element that transforms a plane wave field into a field with two maxima in the intensity distribution profile (the two-lobe field) rotating in the course of propagation. Therefore, within the observation plane the image of the point object looks like two high intensity peaks. The object longitudinal displacement can be evaluated by the change of the intensity maxima position within the observation plane. And the higher the rotation velocity of the intensity of the field generated by the diffraction element is, the more accurate the estimation of the longitudinal coordinate of the emitting object can be. By the rotation velocity we mean the

* Corresponding author: kotova@fian.smr.ru

DOI: 10.1051/epjconf/201713202012

value that determines the rotation angle of the intensity distribution of the light field under its propagation in the longitudinal direction.

The paper is dedicated to the study of spiral-optics-based generation of the phase optical elements that form two-lobe light fields with the intensity distribution rotating at their propagation. Discovered in nineties of the XX-th century, the spiral beams were well-studied [4]. Spiral beams are paraxial light fields with a transverse intensity distribution which rotates around the beam axis maintaining its shape (but not the scale) whilst propagating. It was shown that the spiral beams could be constructed so that to obtain the field with predetermined intensity distribution and rotation parameters. The expression defining the rotation angle θ of the spiral beam intensity in the plane located at z–distance from the initial plane is as follows:

$$\theta(z) = \theta_0 \arctan\left(\frac{2z}{k\rho^2}\right),$$

where k is a wave number ρ – parameter describing the beam lateral dimension, θ_0 – parameter describing the beam rotation velocity during its propagation. A full angle of the beam rotation from the waist to a remote diffraction zone is:

$$\theta = \theta_0 \frac{\pi}{2}.$$

Two approaches were used to obtain phase distributions for the generation of two-lobe fields with different rotation velocity of the intensity distribution. In the first one the initial spiral beam is a sum of several Laguerre-Gauss modes [5], in the second one it was obtained by functional representation [6]. Phase filters were developed for the generation of fields with the rotation parameters $\theta_0 = -1$, $\theta_0 = -2$ and $\theta_0 = -4$. By numerical modeling, parameters of the field illuminating the phase mask were determined for the most effective generation of the light fields under research and were used in experiments conducted.

The experiments on generation of two-lobe light fields were fulfilled with the phase spatial liquid crystal modulator of HOLOEYE HEO-1080 type. Obtained dependencies for the rotation angle vs distance well agreed with the theoretical ones. Note that the energy effectiveness went down with the increase of θ_0 value. For the experimentally obtained light fields with the highest rotation velocity it was not exceeding 30%, while for the case of $\theta_0 = -2$ it achieved its maximum value of 56%.

A number of experiments on evaluation of the effect of coma, astigmatism and spherical aberrations on two-lobe light fields were realized. It was discovered that for the values of 0,198λ, 0,177λ and 0,153λ for spherical aberrations, coma and astigmatism respectively, it was still feasible to get the information on the angular position of the two maxima of the field portions.

The support from Russian Foundation for Basic Research is acknowledged (16-29-11809).

References

1. A.V. Naumov, Phys. Usp. **56**, 605 (2013)
2. A.V. Naumov, I.Yu. Eremchev, A.A. Gorshelev, Eur. Phys. J. D **68**, 348 (2014)
3. S.R.P. Pavani, R. Piestun, Opt. Express **16**, 3484 (2008)
4. E.G. Abramochkin, V.G. Volostnikov, *Modern Optics of Gaussian beams* (Moscow, Fizmatlit, 2010)
5. V.G. Volostnikov, E.N. Vorontsov, S.P. Kotova, N.N. Losevsky, D.V. Prokopova, Bull. Ras: Physics **80**, 766 (2016)
6. E. Razueva, E. Abramochkin, EPJ Web of Conferences **103**, 10011 (2015)

LABS problem and ground state spin glasses system

A.N. Leukhin*, V.I. Bezrodnyi, and Yu.A. Kozlova

Mari State University, 424000 Yoshkar-Ola, Russia

Abstract. In our work we demonstrate the new results of an exhaustive search for optimal binary sequences with minimum peak sidelobe (MPS) up to length N=85. The design problem for law autocorrelation binary sequences (LABS) is a notoriously difficult computational problem which is numbered as the problem number 005 in CSPLib. In statistical physics LABS problem can be interrepted as the energy of N iteracting Ising spins. This is a Bernasconi model. Due to this connection to physics we refer a binary sequence as one-dimensional spin lattice. At this assumption optimal binary sequences by merit factor (MF) criteria are the ground-state spin system without disorder which exhibits a glassy regime.

Design method for ground-state level quantum system for one-dimensional Ising spin chain [1] consisting from N mutual equal-distance particles is offered. Within the frame of Bernasconi model [2] the design problem of simplest quantum system presented like one-dimensional Ising model is reduced to the design problem of binary sequences with lowest level of energy of sidelobes [3,4] of aperiodic autocorrelation. In fact, the Bernasconi model exhibits features of a glass transition like a jump in the specific heat and slow dynammics and aging.

The problem of the glass state remains one major unsolved issued in condensed matter theory. Despite an enormous body of experimental and numerical data and quite detailed phenomenological theories, there is no fully stisfactory microsopic model for the glass state. Within the Bernascni model the high temperature phase of Ising spin system reproduces exactly an approximation due to Golay [3]. For the low-temperature regime, analytical results are rare – especialy for the ground state are not known. Low temperature glass phase is rather similar to the low temperature phase of Derrida's random energy model [15].

There are two criteria for the optimality of binary sequences with low levels of aperiodic autocorrelation: minmum peak sidlobe (PSL) and maximum MF. The effective computing methods for an exhaustive search of binary sequences with maximum MF are presented in [5-7]. Optimal binary sequences by MF criteria are consructed for the lengths N=[2,66]. Enumerative algorihms (complete or partial) are limited to small values of N<300 by the eponential complexity of local search algorithms. Analytical binary sequences for unlimited N with current record of an symptotic merit factor of MF=6.340261 is set called appended rotated Legenfre sequences.

* Corresponding author: leukhinan@list.ru

The effective computing method for an exhaustive search of MPS binary sequences is presented in [8]. Optimal binary sequences by MPS criteria are consructed for the lengths N=[2,85]. The large values of N exhaustive enumeration is not feasible and one has to resort either partial enumerations or heuristic searches. In both cases one obtains sequences with small PSL but not necessarily minimum peaksidelobe. The good compilation of results for lengths N=[2 ;300] and for some lengths between N=303 and N=1000 and for some lengths between N=1019 and N=4096 is given in [16]. The results on adiabatic quantum computer D-Wave 2 for finding low peak sidelobes codes with lengths more than 100 and up to 426 are presented in [9]. The theoretical bound for MPS sequences is offered in [17] and it is given as $PSL \approx 0.435\sqrt{N}$. Some analytical constructions of binary sequences has a PSL between $0.9\sqrt{N} < PSL < 1.1\sqrt{N}$.

We present as some improvements to previously known exhaustive search algorithm for MPS sequences and as some improvements to previously known results of an exhaustive search for MPS [10-13]. An exhaustive search of binary MPS sequences was performed off and on during 32 months using high performance computing grid [14]. Total performance of MarGrid is around 100T/Flops with 120 Intel Xeon E5-2679 v3 processors, with 100 Intel I7 Core processors and with supercomputer Flagman RX240 on the base of 8 NVIDIA TESLA C2059 with 3584 parallel graphical processors and on the base of 2 processors Intel Xeon X5670 (up to Six-Core). For example an exhaustive search for the length $N = 80$ was performed in the background for 1 month. Our computational results are compared with results of some different stochastic local search methods for binary sequences such as: evolutionary algorithm, constrained local search, local search relaxation, direct stochastic search, tabu search, memetic, ants colony optimization, simulated annealing, genetic, iterative local search, scatter search. Executing an exhaustive search for binary sequenceswith PSL=4 we proove that there are no exist any binary sequences with PSL=4 with the lengths N=83, 84, 85.

This work was supported by Russian Ministry of Education and Science (02.G25.31.0204, 2014/81/1856), Russian Foundation of Basic Research (15-07-99514).

References

1. E. Ising, Z. Phys. **31**, 253 (1925)
2. J. Bernasconi, J. Phys. **48**, 559 (1987)
3. M.J.E. Golay, IEEE Trans. Inf. Theory **IT-28**, 543 (1982)
4. CSPLib: a problem library for contraints (www.cslib.org)
5. S. Mertens, J. Phys A: Math. **29**, 473 (1996)
6. H. Bauke, S. Mertens, Phys. Rev. E 70, 025102 (2004)
7. T. Packebusch, S. Mertens, J.Phys. A: Math. Theory. **49**, 165001 (2016)
8. G.E. Coxson, J.C. Russo, IEEE Trans. on Aerosp. and El. Syst. **41**, 302 (2005)
9. G.E. Coxson, C.R. Hill, J.C. Russo, HPEC Conf. **7** (2014)
10. A.N. Leukhin, E.N. Potekhin., Proc. EuRad. 495 (2013)
11. A.N. Leukhin, A.S. Shuvalov, E.N. Potekhin, Bull. RAS. Phys. **78**, 207 (2014)
12. A.N. Leukhin, E.N. Potekhin, Lect. Notes in Comp. Sci. **8865**, 157 (2014)
13. A.N. Leukhin, E.N. Potekhin, Journal of Phys. Conf. Ser. **613**, 012006 (2015)
14. hpc.marsu.ru
15. B. Derrida, Phys. Rev. B **24**, 2613 (1981)
16. W.H. Mow, K.L. Du, W.H. Wu, IEEE Trans. on Aerosp. and El. **51**, 290 (2015)
17. L. Ein-Dor, I. Kanter, W. Kinzel, Phys. Rev. E **65**, 020102 (2002)

Effects of quantum interference in spectra of cascade spontaneous emission from multilevel systems

A.A. Makarov[1,2,*], and V.I. Yudson[1,3]

[1]Institute for Spectroscopy RAS, 108840 Moscow, Troitsk, Russia
[2]Moscow Institute of Physics and Technology, 141700 Dolgoprudny, Russia
[3]Russian Quantum Center, 143025 Skolkovo, Russia

Abstract. A general expression for the spectrum of cascade spontaneous emission from an arbitrary multilevel system is presented. Effects of the quantum interference of photons emitted in different transitions are analyzed. These effects are especially essential when the transition frequencies are close. Several examples are considered: (i) Three-level system; (ii) Harmonic oscillator; (iii) System with equidistant levels and equal rates of the spontaneous decay for all the transitions; (iv) Dicke superradiance model.

The standard textbook expression (see, e.g., Ref. [1]) for the spectrum of spontaneous emission from a (N+1)-level system with the states $|0\rangle, |1\rangle, \dots, |N\rangle$ includes a set of Lorentzians centered at the frequencies $\omega_{N,N-1}$, ... , ω_{21}, and ω_{10}, with their widths $\Gamma_{N,N-1}+\Gamma_{N-1,N-2}$, ... , $\Gamma_{21}+\Gamma_{10}$, and Γ_{10}, respectively. However, if the transition frequencies are close, the Lorentzians are superimposed. This means that the correspondence of a photon to a particular transition is a priori unknown, and this is just the case of the quantum interference. Nevertheless, even for such a more complicated situation the generic formula for the spectrum exists, although its derivation is far from being routine [2]. This exact formula is given by

$$S(\omega)=\frac{1}{2\pi i}\sum_{l=1}^{N}\frac{1}{\Gamma_{l,l-1}}\sum_{k=1}^{l}\prod_{j=k}^{l}\frac{\Gamma_{j,j-1}}{\omega-(E_j-E_{j-1})-\frac{i}{2}(\Gamma_{j,j-1}+\Gamma_{j-1,j-2})}+c.c.\,, \tag{1}$$

where E_j is the energy of the state $|j\rangle$, and the spectrum is normalized as $\int S(\omega)d\omega=N$ by the quantity of photons emitted from $t=0$ to $t\to\infty$ if at the initial moment of time $t=0$ the occupied is the state $|N\rangle$. Below we consider several representative examples of usage of Eq. (1).

Generic three-level system. The spectrum can be presented in terms of the real and imaginary parts of two complex Lorentzians with the widths Γ_{10} and $\Gamma_{21}+\Gamma_{10}$. The weights

* Corresponding author: amakarov@isan.troitsk.ru

of the real parts are $1 \pm 2\Gamma_{21}\Gamma_{10}/(4\Delta^2 + \Gamma_{21}^2)$, and the weights of the imaginary parts are $\mp 4\Gamma_{10}\Delta/(4\Delta^2 + \Gamma_{21}^2)$, where $\Delta = 2E_1 - E_0 - E_2$ is the difference between the frequencies of $|1\rangle \to |0\rangle$ and $|2\rangle \to |1\rangle$ transitions. It can be noticed that at $\Delta = 0$ and $\Gamma_{21} = 2\Gamma_{10}$ the spectrum becomes single-Lorentzian with the width Γ_{10}.

Harmonic oscillator. A generalization of the latter observation is the property of the system with $\Gamma_{j,j-1} = j\Gamma_{10}$, i.e. the harmonic oscillator. No matter whence the oscillator starts - its emission spectrum remains single-Lorentzian with the width Γ_{10} of the lowest transition. This property is, of course, well known (see, e.g., Ref. [3]), but never has been derived using the consistent treatment in terms of the quantum electrodynamics.

Equidistant system with the equal decay rates Γ for all transitions. Here, at the initial excitation of the N-th level, the emission spectrum becomes narrower as the number N grows (see the Figure). Finally, at $N \to \infty$ the spectrum (more precisely, the ratio S=N) degenerates into the δ-function. This observation is in full correspondence with the spectrum of the elastic scattering in a week monochromatic laser field in terms of the quasi-energy approach.

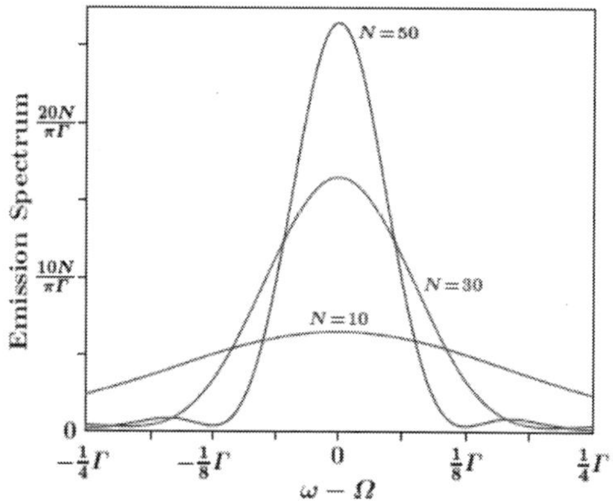

Fig. 1. Emission spectrum (Ω is the frequency of all the successive transitions).

Dicke superradiance model. The Dicke model corresponds to a system of N closely spaced non-interacting two-level atoms. At the initial excitation of all the atoms, a cascade of transitions occurs between the equidistant (Ω is the energy difference) collective states (Dicke states). The emission spectrum calculated with the use of the Bethe Ansatz [4] coincides with that given by the general formula (1) where the transition rates are $\Gamma_{j,j-1} = j(N-j+1)\Gamma$. When $N \gg 1$, the spectrum takes the form

$$S(\omega) = \frac{1}{2\pi\Gamma}\frac{1}{\cosh^2(\pi\varpi)}, \quad \varpi = \frac{\omega - \Omega}{N\Gamma} \tag{2}$$

The work was supported by Russian Foundation for Basic Research (No. 15-02-05657).

References

1. V.B. Berestetskii, E.M. Lifshitz, L.P. Pitaevskii, *Quantum Electrodynamics, Course of Theoretical Physics Vol. 4* (Pergamon, Oxford, 1982)
2. A.A. Makarov, V.I. Yudson, Phys. Rev. A **89**, 053806 (2014)
3. B.Ya. Zeldovich, A.M. Perelomov, V.S. Popov, Sov. Phys. JETP **28**, 308 (1969)
4. V.I. Rupasov, V.I. Yudson, Sov. Phys. JETP **60**, 927 (1984)

Defects in regular nanosystems and interference spectra at reemission of electromagnetic field attosecond pulses

V.I. Matveev, and D.N. Makarov*

M.V. Lomonosov Northern (Arctic) Federal University, 163002 Arkhangelsk, Russia

Abstract. The effect of defects in nanostructured targets on interference spectra at the reemission of attosecond electromagnetic pulses has been considered. General expressions have been obtained for calculations of spectral distributions for one-, two-, and three-dimensional multiatomic nanosystems consisting of identical complex atoms with defects such as bends, vacancies, and breaks. Changes in interference spectra by a linear chain with several removed atoms (chain with breaks) and by a linear chain with a bend have been calculated as examples allowing a simple analytical representation. Generalization to two- and three-dimensional nanosystems has been developed.

Crystals and nanostructured targets are natural diffraction gratings for X-rays. The diffraction of X-rays by various periodic structures is usually described as the scattering of plane waves with infinite time duration [1]. The scattering of attosecond electromagnetic pulses from such structures has been poorly studied to date. At the same time, such processes can supplement the X-ray diffraction analysis by the capabilities of high time resolution spectroscopy, including attosecond spectroscopy and attosecond metrology [2–6]. Nevertheless, the processes of interference at the scattering of attosecond pulses from various regular targets have been studied only in a few works. The theory of the reemission of attosecond electromagnetic pulses by arbitrary multiatomic systems consisting of isolated complex atoms was developed in [7–9]. Interference effects become dominant in the rescattering spectra when targets are regular structures with a large number of atoms. In this case, the angular distributions of the incident and scattered radiations can be significantly separated by choosing the spatial structure of targets and their various combinations. The theory of the reemission of attosecond electromagnetic pulses by arbitrary regular multiatomic systems consisting of identical complex atoms was developed in [10] with allowance for chaotic thermal vibrations. However, interference processes at the reemission of attosecond electromagnetic pulses by arbitrary regular multiatomic systems containing various defects have not yet been considered.

In this work, the effect of defects in nanostructured targets on interference spectra at the reemission of attosecond electromagnetic pulses is considered. General expressions were obtained for calculations of spectral distributions for one-, two-, and three-dimensional

* Corresponding author: d.makarov@narfu.ru

multiatomic nanosystems consisting of identical complex atoms with defects such as bends, vacancies, and breaks. Changes in interference spectra by a linear chain with several removed atoms (chain with breaks) and by a linear chain with a bend have been calculated as examples allowing a simple analytical representation. Generalization to two- and three-dimensional nanosystems has been developed. The proposed approach can be directly expanded to more general types of defects. It is noteworthy that the results can pertain not only to one-dimensional and rectangular two- and three-dimensional regular nanostructures. The interference factors of various regular two- and three-dimensional nanostructures determined in [8,9] for planar and cylindrical structures can be used for models of various nanosystems (rings and nanotubes, including multilayer nanotubes and "forest" of nanotubes). In the cases under consideration, it is assumed that the duration of attosecond pulses is much smaller than the characteristic atomic time. In this work, as in [7–10], we consider of the emission of one photon by all atoms of the target in the time of action of sudden perturbation. After the action of sudden perturbation, excited atoms of the target can relax with the emission of photons belonging to the known spectra of isolated atoms. However, if sudden perturbation changes the velocities of atomic electrons, atoms can emit even during its action [11]. A classical analog of the problem in this formulation is the known example [12] of the spectrum of radiation of a free electron at a sudden change of the velocity. After sudden perturbation, atoms of the target are distributed over various excited states. In this case, they can relax with the emission of photons (in time characteristic of radiative transitions). Interference effects will obviously be absent in this case. Furthermore, reemission spectra and spectra emitted at relaxation are strictly separated in time. The reemission spectrum is emitted only in the time of action of an attosecond pulse, whereas the relaxation spectrum is emitted after the action of the attosecond pulse. Thus, the reemission spectra can be identified in the scheme of coincidences with the attosecond pulse.

This work was supported by the Ministry of Education and Science of the Russian Federation (state assignment no. 3.1726.2014/K) and in part by the Russian Foundation for Basic Research (project no. 15-02-01894) and by the Council of the President of the Russian Federation for Support of Young Scientists and Leading Scientific Schools (project no. SP-1800.2015.1).

References

1. J.M. Cowley, *Diffraction Physics* (North-Holland, Amsterdam, 1975)
2. P. Agostini, Rep. Prog. Phys. **67**, 813 (2004)
3. P.V. Corkit, F. Krausz, Nature Phys. **3**, 381 (2007)
4. F. Krausz, M. Ivanov, Rev. Mod. Phys. **81**, 163 (2009)
5. A.M. Zheltikov, Phys. Usp. **54**, 29 (2011)
6. J. Fink, E. Schierle, E. Weschke, J. Geck, Rep. Prog. Phys. **76**, 056502 (2013)
7. V.I. Matveev, D.U. Matrasulov, JETP Lett. **96**, 628 (2012)
8. D.N. Makarov, V.I. Matveev, J. Exp. Theor. Phys. **117**, 784 (2013)
9. D.N. Makarov, V.I. Matveev, Opt. Spectrosc. **116**, 163 (2014)
10. D.N. Makarov, V.I. Matveev, JETP Lett. **101**, 603 (2015)
11. V.I. Matveev, J. Exp. Theor. Phys. **97**, 915 (2003)
12. L.D. Landau, E.M. Lifshitz, *Course of Theoretical Physics, Vol. 2: The Classical Theory of Fields* (Pergamon, Oxford, 1975)

Optical spectroscopy of random deformations in elastically-anisotropic crystals containing rare-earth ions

B.Z. Malkin[1,*], E.I. Baibekov[1], N.M. Abishev[1], D.S. Pytalev[2], M.N. Popova[2], and M. Bettinelli[3]

[1]Kazan Federal University, 420008 Kazan, Russia
[2]Institute for Spectroscopy RAS, 108840 Moscow, Troitsk, Russia
[3]Dipartimento di Biotecnologie, University of Verona and INSTM, 37134 Verona, Italy

Abstract. We present the results of studies of spectral effects in the optical high-resolution (0.01 cm-1) spectra of rare-earth ions in crystals caused by random deformations of a crystal lattice. Low-temperature polarized transmission spectra in a broad spectral range (5000–15000 cm^{-1}) were taken for tetragonal single crystals ABO4 (A=Y, Lu; B=V, P) containing impurity Tm^{3+} ions with concentrations 0.2 and 1.0 at.%. A specific fine structure of singlet-doublet transitions in the Tm3+ ions was observed. We demonstrate a possibility to estimate a concentration of intrinsic lattice defects from the analysis of the measurement data, by making use of an analytical expression derived in the present work for the distribution function of random lattice strains induced by point defects in the elastically-anisotropic continuum.

Modulation of crystal fields by random lattice deformations brings inhomogeneous broadening of spectral lines. Random deformations are also responsible for a specific fine structure of transitions which involve degenerate crystal-field sublevels of rare-earth ions in the crystal fields of cubic, tetragonal, or trigonal symmetry [1-3].

In the present work, we measured polarized high-resolution (0.01 cm^{-1}) transmission spectra of RVO_4 and RPO_4 (R=Y, Lu) crystals with the zircon structure containing impurity Tm^{3+} ions (0.2 and 1 at.%) which substitute for the R^{3+} ions at sites with the local tetragonal D_{2d} symmetry. At temperatures below 15 K, we observed a doublet structure of the spectral lines corresponding to transitions between the singlet (doublet) and doublet (singlet) crystal-field sublevels of the ground multiplet $^3H_6$ and excited multiplets $^3F_4$, $^3H_5$ and $^3H_4$, respectively, of the Tm^{3+} ions. These lines have a specific shape (see Fig. 1) with a narrow (as compared with a full line width at half maximum) dip at the center and the splittings between the maxima from 0.04 up to 0.16 cm^{-1} for different lines, increasing with the concentration of the tulium ions and markedly exceeding hyperfine splittings of the doublets.

Calculations of crystal-field energies of the Tm^{3+} ions in the deformed crystal lattice were carried out using the Hamiltonian that involved energies of 4f-electrons in the free

* Corresponding author: boris.malkin@kpfu.ru

ion, hyperfine interactions, crystal-field and electron-deformation interactions. Parameters of the electron-deformation interaction were calculated in the framework of the exchange-charge model and corrected by the analysis of the piezo-spectroscopic measurements [4].

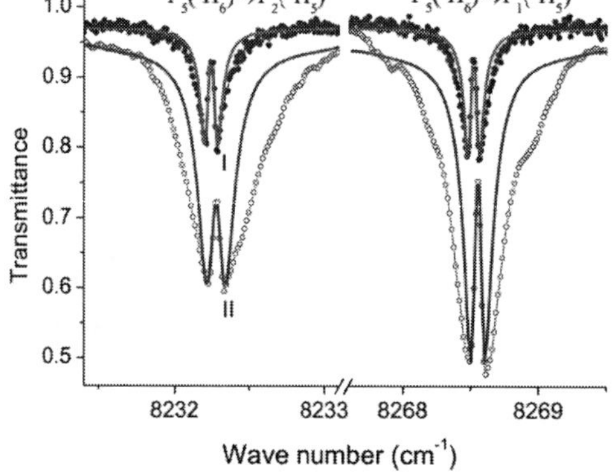

Fig. 1. Fragments of the measured transmission spectra of YVO_4 single crystals doped with Tm^{3+} ions (0.2 at.% (I) and 1.0 at.% (II)) at T=4 K (symbols). Solid lines represent results of calculations.

The envelopes of the spectral lines were obtained by averaging transition intensities calculated for fixed values of the strain tensor components with the distribution function of strains induced by point defects in the elastically-anisotropic continuum. The derived analytic expression for the distribution function presents the generalized Lorentz distribution in the six-fold space of the strain tensor components e_α $(\alpha = 1:6)$: $g(\boldsymbol{e}) = (15\gamma / 8\pi^3)\det|v_{\alpha\beta}|\sum_{\alpha,\beta=1}^{6}(v_{\alpha\beta}e_\alpha e_\beta + \gamma^2)^{-7/2}$. Here, $v_{\alpha\beta}$ are coefficients in the positively determined quadratic form constructed from the strain tensor components. Values of these coefficients were found using Green's functions of the elastic theory equations [5] computed for the sets of elastic constants of the studied crystals. Examples of the calculated envelopes of the spectral lines are shown in Fig. 1. As follows from a comparison of the computed and measured spectra, interactions between the impurity ions contribute markedly into the line widths even in crystals with relatively small concentrations (of about 1%) of impurity ions.

When both concentrations of the intrinsic point lattice defects (C_i) and of the impurity ions (C_{Tm}) are small enough, the width γ of the distribution function is a linear function of C_i and C_{Tm}. Using the widths γ obtained from the analysis of the measured line shapes of different singlet-doublet transitions in crystals with concentrations of 0.2 and 1 at.% of the Tm^{3+} ions, we estimated concentrations of intrinsic lattice defects. Even in the crystals of high optical quality, the width of the random strains distribution, corresponding to the intrinsic defects, is not less than (3 – 5) 10^{-5}. The results of the present work open a possibility to connect the observed spectral effects with the defect «strength» and may serve as a basis for the elaboration of the quantitative method of crystal quality control.

This study was supported by RAS in the frame of the Programs for Basic Research.

References

1. S.A. Klimin et al., Phys. Rev. B **81**, 045113 (2010)
2. B.Z. Malkin et al., Phys. Rev. B **86**, 134110 (2012)
3. G.S. Shakurov et al., Phys. Chem. Chem. Phys. **16**, 24727 (2014)
4. A.V. Vinokurov et al., Phys. Solid State **30**, 2415 (1988)
5. I.M. Lifshits, L.N. Rozentsvaig, JETP **17**, 783 (1947)

DOI: 10.1051/epjconf/201713202017

The polarization approach to registration femtosecond time slots based stimulated photon echo on the exciton states

I.I. Popov[1,*], N.S. Vashurin[1], S.E. Putilin[2], E.A. Vinogradov[3], and A.U. Bahodur[1]

[1]Volga State University of Technology, 424000 Yoshkar-Ola, Russia
[2]Information Technologies, Mechanics and Optics University, 197101 Saint-Petersburg, Russia
[3]Institute for Spectroscopy RAS, 108840 Troitsk, Moscow, Russia

Abstract. It is reported about the first experiments on the registration of a femtosecond time interval using the effect of non-Faraday rotation of the plane polarization of the stimulated photon echo. The experiment was performed at room temperature in a three-layer semiconductor film thickness of 300 nm, obtained by magnetron sputtering. The fundamental difference of this effect from the Faraday's effect was shown. We give conclusions about the possibility of femtosecond time intervals registration at room temperature with an accuracy of 25 fs in the polarization principle of action, rather than spectral.

The aim of the work is aimed at demonstrating the possibility of building on the basis of the effect of the new femtosecond magneto-optics principle of femtosecond time intervals registration. This principle of operation is used polarizing approach to fixing the value of these slotsTo achieve this goal have been made the work for the detection [1] and for quantitative assessment of non-Faraday rotation of the polarization plane of femtosecond stimulated photon echo. This femtosecond stimulated photon echo was observed at room temperature on exciton states of thin textured semiconductor film.

The research method is based on the polarization properties of a stimulated photon echo. Resonance medium - is textured three-layer semiconductor film ZnO/Si(B)/Si(P) (each layer having a thickness of 100 nm) obtained by magnetron sputtering. The film consists of crystalline fibers oriented orthogonally to the film surface. The distance between the fibers is filled with X-ray amorphous phase sprayed substances. Crystal fiber with distance from the substrate decreased the transverse dimensions. This led to the emergence surface nanoscale crystal defects. Open (deformation) of the crystal structure of the fibers at the border of contact with the X-ray amorphous phasecreates conditions for the localization of excitons on these surface defects and creates conditions of inhomogeneous broadening of the excitonic spectral lines.

In theory, this effect is predicted for the gas atmospheres in the work [3] The first experiment on non-Faraday rotation of the plane polarization of the stimulated photon echo made by the authors of [4] Details polarization dependence of this effect is studied in [5] For the first time this effect on exciton gas detected experimentally in [1, 2]. The proposed polarization approach to the registration ultrashort time intervals of at least an order of magnitude superior to existing methods based on the spectrum approach, using devices

* Corresponding author: popov@volgatech.net

such as smiling in the "Pikohron" with a resolution of up to 0.5 ps. Discussed in this paper, the effect, in contrast to the rotation of the polarization plane in optically active (chiral) media and the Faraday rotation of the plane of polarization does not depend on the concentration of the growth of the active particles and the length of the path traversed by the radiation in resonance mediumAt the same time, the angle of rotation of the polarization plane of the echo signal φ defines by type of quantum transition and magnetic field produced by the amplitude and time delay between the exciting laser pulses. Conducted and the rank of the angle φ exceeds the value of the Faraday rotation by 3-4 orders of magnitude (10^3 – 10^4). The authors of [4] it is shown that, depending on the type of transition, degenerate in the magnetic quantum number, (J ← for molecules transitions (this corresponds to the spectral lines P -, R or Q-type) for line P -, R-type, this rotation is of the order of degrees, and for Q-type lines, respectively, 90 degrees. This corresponds to the theoretical predictions of this effect .

One of the prerequisites for the emergence of the non-Faraday effect rotation of the plane of polarization of the photon echo is different splitting in a magnetic field of the upper and lower levels of the quantum transition [6]. If we neglect the influence of the exciton-phonon interaction on the formation of the echo signal, taking into account this error occurs when using the analytical expressions for gas at a fixed value of the longitudinal magnetic field component in our experiments, 0.4 Tesla, we can estimate the degree of dependence of the angle SPE polarization plane rotation φ of the time interval between the second and third exciting pulses. This relationship will have a form of the function $\varphi = f(\text{arctg}\ \tau_{23})$. To exclude a contribution to the turnaround time interval $\tau_{12,}$ it is assumed to be zero. Figure 1 shows the discovery non-Faraday SPE rotation of the plane of polarization in the thin-layer film and its relationship to the magnitude of the time interval between the second and third laser pulses, the excitation signal photon echo.

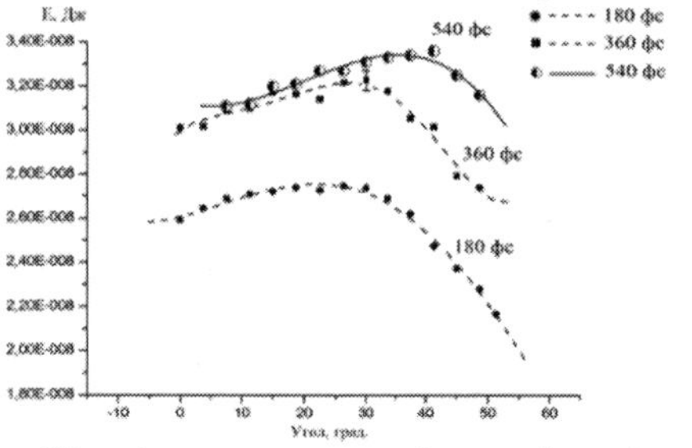

Fig. 1. Dependence non-Faraday rotation of the plane of polarization of the stimulated photon echo on the value of the time interval between the second and the third exciting pulses: at τ_{23} 180 fs, the rotation angle was 25 degrees, with 360 fs - 32 degrees, at 540 fs - 40 degrees.

From Fig. 1, one can see that registration by changing the angle φ, consisting of 7 or 8 degrees allows to determine the time interval $\tau_{23,}$ amounted to 180 fs. Using non-Faraday rotation of the plane of polarization of the stimulated photon echo one can determine the magnitude of the time intervals of 180 fs to 25 fs accuracy of registration degree.

Unusual behavior of the dysprosium CF energies in Dy_2BaNiO was detected in the temperature range lower than T_N. All CF levels experience a shift (up to 10 cm^{-1}) from their positions at T_N (see, e.g., Fig. 1). Such behavior can be explained as being due to the spontaneous electric polarization that appears in the temperature range discussed [8]. Redistribution of the charge density influences the crystal field acting on dysprosium and leads to a renormalization of CF self-energies.

References

1. I.I. Popov et al., *Proc. of 10th Int. Sci. Sch. "Sci. & Innovations"* (Yoshkar-Ola, 2015)
2. I.I. Popov, Memoirs of the Faculty of Physics Moscow St. Univ. **4**, 154338 (2015)
3. I.V. Evseev, V.M. Ermachenko, V.A Rechetov, Opt. Spectrosc. **52**, 444 (1982)
4. I.S. Bikbov, I.I. Popov, V.V. Samartsev, I.V. Yevseyev, Las. Phys. **5**, 580 (1995)
5. I.V. Yevseyev et al., Las. Phys. **16**, 535 (2006)
6. A.I. Alekseev, JETP Lett. **9**, 472 (1969)

Features of surface enhanced Raman scattering in the systems with «hot spots»

E.V. Solovyeva*, D.A. Khazieva, and A.S. Denisova

Saint-Petersburg State University, Institute of Chemistry, 199034 Saint-Petersburg, Russia

Abstract. In this work we demonstrate the features of SERS on the substrates with «hot spots» on the example of system «diaminostilbene - colloidal silver». We found that «hot spots» forming on aggregated nanoparticles exist on the metal substrates only at low concentration of ligand. This effect caused by the gradual filling of first monolayer by adsorbate molecules. Significantly higher enhancement factor is obtained for substrates with «hot spots», for which the participation of resonance processes in the formation of SERS signal is revealed also.

Surface enhanced Raman scattering (SERS) is the rapidly developing technique of vibrational spectroscopy based on the phenomenon of Raman signal amplification for molecules adsorbed on metallic nanostructured surface or nanoparticles. Recently, the special interest was devoted to the SERS investigation on substrates which have "hot spots" - areas where the surface plasmons are localized [1]. One consequence of this effect is more significant amplification of optical signal from the molecules placed in the "hot spots", that is especially valuable for vibrational spectroscopy, in particular, for SERS [2].

The presence of "hot spots" is typical for substrates with closely spaced nanoparticles where the interparticle gap does not exceed a few nanometers. One of the methods for obtaining such substrates is linking of nanoparticles by chemical modifiers. Usually, the modifiers are compounds with two spatially separated functional groups, where each of them is capable to react with metal surface. In this work we have shown the features of SERS on the substrates with "hot spots" obtained by linking silver nanoparticles via 4,4'-diaminostilbene (DAS) molecules. DAS has conjugated system and two amino groups located on the opposite ends. Due to this structure DAS is attractive for creation the substrates with "hot spots".

For SERS spectra of DAS it was found that their intensity changes in non-monotonic way with the increasing of ligand concentration. Also, the profile of SERS spectra transforms dramatically at concentration equal to $5\cdot10^{-5}$ M (Fig. 1).

The observed behavior of DAS SERS spectra is caused by two main factors associated with the gradual filling of surface area by adsorbate molecules. The one of them is "the effect of first monolayer" when the greatest enhancement of Raman signal arises for molecules from the first monolayer. The next is the existence of "hot spots" in silver colloid solely at low concentration of DAS. In this case the partial coverage of the surface is

* Corresponding author: solovyeva_elena_v@gmail.com

reached only. The last supposition was confirmed by scanning electron microscopy and UV-VIS spectroscopy.

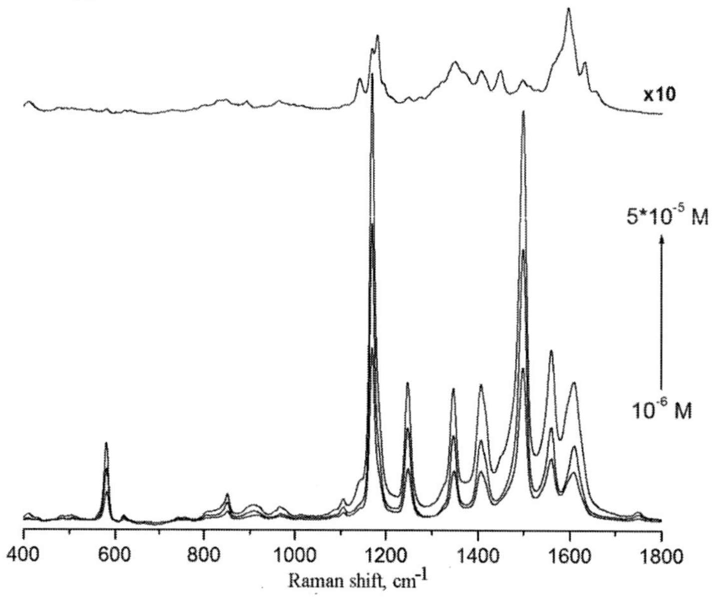

Fig. 1. SERS spectra of DAS adsorbed on silver nanoparticles at various concentrations; the spectrum at $C_{DAS} = 5 \cdot 10^{-5}$ M is multiplied by 10 and shifted along axis y.

This work was supported by the research Grant (number 12.38.235.2014) of Saint-Petersburg State University. The authors acknowledge the Research Center for X-ray Diffraction Studies and Nanotechnology Interdisciplinary Center of Saint-Petersburg State University.

References

1. J. Kubackova, I. Izquierdo-Lorenzo, D. Jancura, P. Miskovsky, S. Sachez-Cortes, Phys. Chem. Chem. Phys. **16**, 11461 (2014)
2. E.Le Ru, P. Etchegoin, *Principles of Surface-Enhanced Raman Spectroscopy and related plasmonic effects* (Amsterdam, Elsevier, 2009)

Wavelength dependence of the efficiency of singlet oxygen generation upon photoexcitation of photosensitizers

A. Starukhin[1,*], V. Knyukshto[1], V. Plavskii[1], A. Gorski[2], and J. Solarski[2]

[1]B.I. Stepanov Institute of Physics, NAS of Belarus, Minsk, Belarus
[2]Institute of Physical Chemistry, Polish Academy of Sciences, Warsaw, Poland

Abstract. The dependence of the efficiency of singlet oxygen ($^1\Delta_g$) generation upon excitation of photosensitizer at different wavelength was observed for several derivatives of palladium porphyrin in carbon tetrachloride. The efficiency of singlet oxygen generation upon excitation in a blue region of the spectrum (Soret band) exceeds by several times the efficiency at excitation in the red spectral region (Q band). The effect of enhancement of singlet oxygen generation upon CW photoexcitation to Soret band of photosensitizer may be explained by influence of high laying triplet states of a donor molecule on the triplet-triplet energy transfer.

Metallocomplexes of porphyrins have been widely used as photosensitizers for the singlet oxygen generation during more than 40 years [see 1.2 and references therein]. The singlet oxygen generation in this case occurs by means of triplet-triplet energy transfer between primarily populated triplet state of photosensitizer and the ground state of molecular oxygen (triplet state - $^3\Sigma^-_g$). The singlet oxygen luminescence has been often used for determination of efficiency of singlet oxygen generation. In many cases, a monochromatic laser light at the visible spectral range has been used as a source of population of the triplet state of photosensitizer. It seems quite interesting to study how the efficiency of singlet oxygen generation depend on wavelength of photoexcitation. The detection of excitation spectra of singlet oxygen luminescence at a whole spectral range of absorption of photosensitizer looks as an appropriate tool for this goal. The method allows to determine the spectral ranges in the absorption spectrum of the sensitizer with the most efficient singlet oxygen generation.

Two peripherally substituted derivatives of porphyrin: Pd-2,3,7,8,12,13,17,18-octaethylporphyrin and Pd-5,10,15,20-tetraphenylporphyrin were investigated (Fig. 1).

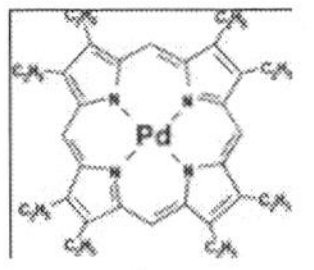

1

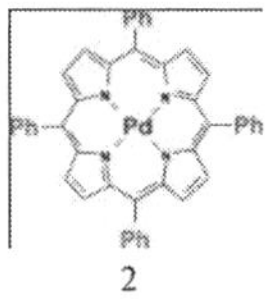

2

Fig. 1. Structure of Pd-OEP(**1**) and Pd-TPP(**2**)

* Corresponding author: a_starukhin@mail.ru

Fluorescence, phosphorescence and fluorescence excitation spectra as well as excitation spectra of Pd-TPP measured at wavelength of singlet oxygen luminescence ($\lambda_{rec.}$=1274 nm) have been recorded at ambient temperature and normal pressure in carbon tetrachloride solution (Fig. 2).

As is evident from Fig. 2, the excitation spectrum of singlet oxygen luminescence for Pd-TPP photosensitizer (1) has significantly higher intensity in the region of the Soret band in comparison to the Pd-TPP fluorescence excitation spectrum (2) which, at the same time, is well correlated with

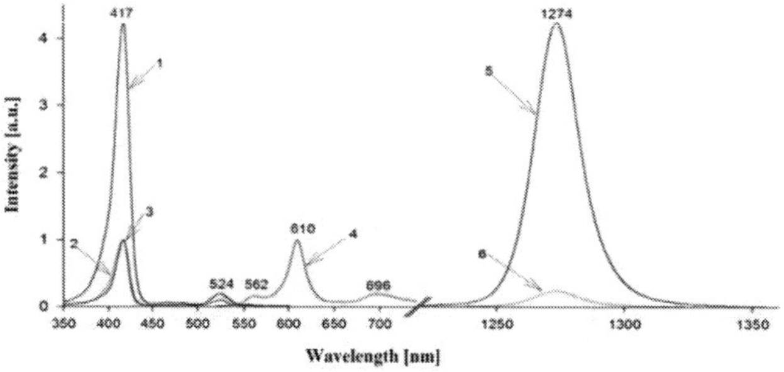

Fig. 2. Excitation spectrum of Pd-TPP at $\lambda_{rec.}$=1274 nm (1), fluorescence excitation spectra at λrec.=610 nm (2), absorption spectrum (3), fluorescence spectrum at $\lambda_{exc.}$=417 nm (4), luminescence spectra of singlet oxygen excited at 417 nm (5) and 524 nm (6)

Pd-TPP absorption spectrum (3). The experimental data allows to conclude that the efficiency of singlet oxygen generation in the case of Pd-TPP photosensitizer is significantly higher for excitation in the Soret band ($\lambda_{exc.}$= 417 nm) than for excitation in the visible region of the spectrum - Q band ($\lambda_{exc.}$=562 nm). Previously, the similar behavior of excitation spectra of singlet oxygen luminescence measured in spectral range of absorption of photosensitizer were reported in [3,4] for other porphyrin derivatives, but the origin of this effect was not well explained.

Photochemical generation of singlet oxygen occurs in result of a triplet-triplet energy transfer between the excited lowest triplet state of photosensitizer and the ground (triplet) state of molecular oxygen. The Pd-TPP exhibits a very strong and wide triplet-triplet absorption band centered at 462 nm (Fig. 3). Taking into account a continuous wave excitation used during the above-mentioned experiment, as well as, an overlap of that strong triplet-triplet absorption band with the ground state absorption in the region of Soret band, an influence of higher triplet states of Pd-TPP on singlet oxygen generation can be postulated. To answer what is the exact origin of the dependency of singlet oxygen generation on wavelength of excitation of photosensitizer an additional experiments needed.

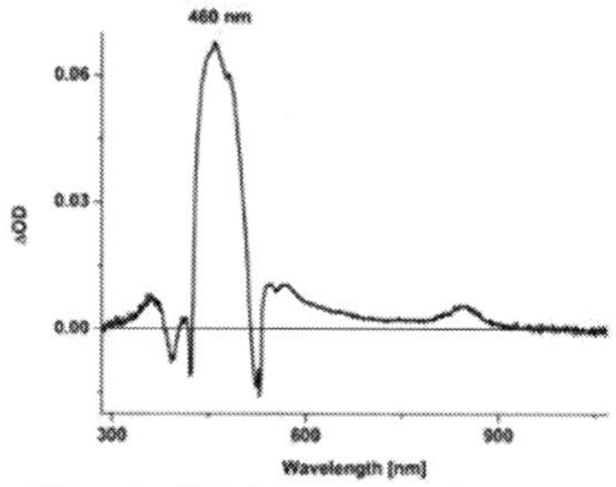

Fig. 3. Triplet-triplet absorption spectrum of Pd-TPP.

The research leading to these results has received funding from the European Union's Horizon 2020 research and innovation programme under grant agreement No 645628.

References

1. A. Krasnovsky jr., Biofizika USSR **21**, 748 (1976)
2. F. Wilkinson, W.Phillip, J. Phys. Chem. Ref. Data Rev. **22** 113 (1993)
3. P. S,ena, C. Hirelb, A.-G. Gürek, C. Andrauda, Y. Bretonnière, M. Lindgrenc, J. Porphyrins Phthalocyanines **17**, 964 (2013)
4. D. Kuznetsov, Yu. Myagchenko, O. Slobodyanyuk, Ukr. J. Phys. **59**, 215 (2014)

Experiment and computation: a combined approach to study the van der Waals complexes

L.A. Surin*

Institute for Spectroscopy RAS, 108840 Troitsk, Moscow, Russia

Abstract. A review of recent results on the millimetre-wave spectroscopy of weakly bound van der Waals complexes, mostly those which contain H_2 and He, is presented. In our work, we compared the experimental spectra to the theoretical bound state results, thus providing a critical test of the quality of the M–H_2 and M–He potential energy surfaces (PESs) which are a key issue for reliable computations of the collisional excitation and de-excitation of molecules (M = CO, NH_3, H_2O) in the dense interstellar medium. The intermolecular interactions with He and H_2 play also an important role for high resolution spectroscopy of helium or *para*-hydrogen clusters doped by a probe molecule (CO, HCN). Such experiments are directed on the detection of superfluid response of molecular rotation in the He and p-H_2 clusters.

The presence of molecular species like CO, H_2O, NH_3 in the interstellar medium (ISM) and their usefulness as indicators of molecular cloud properties has motivated considerable experimental and theoretical efforts aimed largely at understanding the interactions between these species and their main ISM collision partners, H_2 and He [1, 2]. The collisional excitations of the species involved strongly depend on important features of the multidimensional intermolecular potential energy surfaces (PES), such as barriers, anisotropies, binding energy, and structure in the most stable configurations.

High resolution spectroscopy of van der Waals complexes has proofed to be a very effective tool for the elucidation of intermolecular PES, because the bound states of the complexes are sensitive to the interaction potential [3]. The weakly bound complexes can be efficiently produced in a supersonic jet expansion of a gas mixture into vacuum. A difficulty in undertaking the spectral measurements of such systems is that the transitions are extremely weak. The recent progress on the millimeter-wave studies of weakly bound molecular complexes, mostly those which contain H_2 and He, using intracavity OROTRON jet spectrometer [4] along with double resonance technique is presented. These experimental measurements provided rigorous tests of the accuracy of the *ab initio* PESs and computed bound states [5, 6].

Another application, in which the intermolecular interactions with He and H_2 play an important role, is the doping of helium or *para*-hydrogen clusters with a probe molecule to perform high resolution spectroscopy. Such experiments are directed on the detection of superfluid response of molecular rotation in the He and p-H_2 clusters [7]. It was shown

* Corresponding author: surin@isan.troitsk.ru

recently that the lighter probe molecules (CO, HCN) allow one to measure the effective inertia of He and *p*-H_2 clusters while maintaining a maximum superfluid response with respect to dopant rotation.

References

1. S. Green, P. Thaddeus, Astrophys. J. **205**, 766 (1976)
2. S. Chefdeville, T. Stoecklin, C. Naulin, P. Jankowski, K. Szalewicz, A. Faure, M. Costes, A. Bergeat, Astrophys. J. Lett. **799**, L9 (2015)
3. P.E.S. Wormer, A. van der Avoird, Chem. Rev. **100**, 4109 (2000)
4. L.A. Surin, B.S. Dumesh, F. Lewen, D.A. Roth, V.P. Kostromin, F.S. Rusin, G. Winnewisser, I. Pak, Rev. Sci. Instrum. **72**, 2535 (2001)
5. P. Jankowski, L. Surin, A. Potapov, S. Schlemmer, A.R.W. McKellar, K. Szalewicz, J. Chem. Phys. **138**, 084307 (2013)
6. A.V. Potapov, L.A. Surin, V.A. Panfilov, B.S. Dumesh, T.F. Giesen, S. Schlemmer, P.L. Raston, W. Jäger, Astrophys. J. **703**, 2108 (2009)
7. L.A. Surin, JETP Lett. **97**, 57 (2013)

Frustration of total internal reflection by a hidden nanowire

Leonid Frumin[1,2], *Anton* Nemykin[1,2], and *David* Shapiro[1,2,*]

[1]Institute of Automation and Electrometry, SB RAS, 630090 Novosibirsk, Russia
[2]Novosibirsk State University, 630090 Novosibirsk, Russia

Abstract. A metallic wire embedded in the dielectric substrate leads to frustration of the total internal reflection at the interface between dielectric and free space. The scattering by a gold cylinder is studied in order to exploit the light for hidden objects optical diagnostics. The p-wave incidenting by the angle close to total reflection with the frequency near plasmonic resonance is treated by the modified boundary elements method. The magnetic and electric fields at the interface are calculated. The scattering diagrams are found in near and far field.

The scattering of evanescent wave by particles is important elementary process of near-field optics [1, 2]. It is a particularly promising method in near-field scanning microscopy [3–5]. The interesting case of metallic particle embedded in dielectric half-space has a number of specific technical applications, for instance, the tomographic reconstruction of hidden scatterer, see [6] and references therein. As distinct from a running wave, the evanescent wave is not a limited solution of Maxwell equations in the whole space. That is why its scattering should be considered along with its source, in our problem the dielectric half-space with incident and reflecting waves. For calculation we exploit the modified boundary elements method (BEM) [7]. The feature of modified BEM is the special Green function exactly taking into account the layered medium (free and dielectric half-spaces). The substitution of Green function makes it possible to use the same numerical scheme as for a running wave. The discretization of infinite boundary is not necessary. Multiple scattering and reflections are accounted automatically. Moreover, the modified BEM has a number of advantages compared to other common methods based on the finite differences or finite elements. BEM reduces the problem to calculation of field at the boundary, then decrements, in fact, the dimension of problem. In our case two-dimension problem reduces to one-dimension boundary integral equations.

In previous work [7, 8] the object is located on the substrate. In present paper the scatterer hidden inside the dielectric is considered. The object is illuminated by the incident wave near the angle close to total internal reflection (TIR). In this case TIR is frustrated by the scattering and strongly depends on the scatterer parameters. The goal is to study the angular diagram of scattering.

The p-wave is considered where the magnetic field has one component along the axis of cylinder. The calculations are performed for gold cylinder with radius 150, 100, 75, 50 nm

* Corresponding author: shapiro@iae.nsk.su

at distance 50, 100 and 150 nm from the interface between glass (dielectric permittivity 2.25) and free space. The
incidence angle θ is close to TIR angle 41.8^0 at the chosen wavelength $\lambda = 1.512\ \mu m$. Fig. 1 illustrates results for cylinder of radius 150 nm at distance 50 nm.

Fig. 1(a) shows the angular diagram of the energy flux of scattered wave, averaged over a period, at the circles with different radius R at $\theta = 41^0$. The major part of radiation is reflected back into the glass with angles $180 - 360^0$. Only a small part of light is scattered into free space $0 - 180^0$. In free space the main part enters the first quadrant $0 - 90^0$, Fig. 1(b). Above the TIR angle indicatrix swings to the second quadrant $90 - 180^0$ and the "backward scattering" prevails, Fig. 1(c), due to the scattering of wave reflected from the plane.

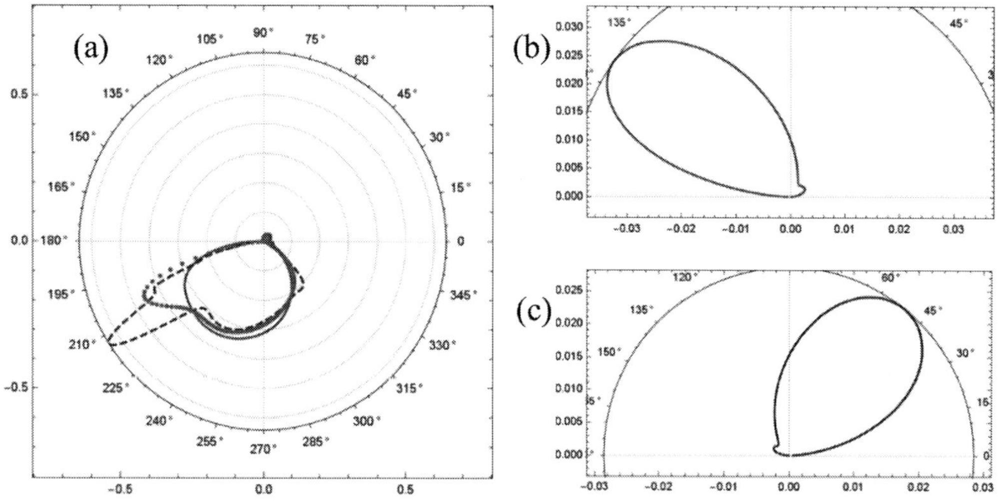

Fig. 1. (a) Scattering indicatrix at $R/\lambda = 1$ (solid curve), 2.5 (dots), 10 (dashes). The magnified angular diagram in the free space at $R/\lambda = 10$ (far field) and $\theta = 41^0$ (b), 43^0 (c).

The signal of frustration in the upper half-plane, where a detector is supposed, turns to be sufficiently weak. There is a way to increase the scattering using a periodic grid instead of single wire. This case can be analyzed by the effective Green function for periodic structure [8].

The Russian Foundation for Basic Research (16-02-00511) and Government program of the leading research schools (NSh-6898.2016.2) support this work.

References

1. L. Novotny, B. Hecht, *Principles of Nano-Optics* (Cambridge, 2006)
2. M.I. Stockman, Phys. Today **64**, 39 (2011)
3. B. Hecht, B. Sick et al., J. Chem. Phys. **112**, 7761 (2000)
4. C. Girard, C. Joachim, S. Gauthier, Rep. Prog. Phys. **63**, 893 (2000)
5. P.S. Carney, J.C. Schotland, Opt. Lett. **26**, 1072 (2001)
6. T. Yamaoki, H. Hamada, O. Matoba, Appl. Opt. **55**, 6874 (2016)
7. O.V. Belai, L.L. Frumin et al., Opt. Lett. **36**, 954 (2011)
8. L.L. Frumin, A.V. Nemykin et al., J. Opt. **15**, 085002 (2013)

EPJ Web of Conferences **132**, 02022 (2017) DOI: 10.1051/epjconf/201713202022
SPECTROSCOPY.SU 2016

Diagnostics of high-temperature plasmas by the X-ray spectra of heavy elements

S.N. Andreev, A.P. Shevelko*, I.Yu. Tolstikhina, L.A. Vainstein, and O.F. Yakushev

P.N. Lebedev Physical Institute RAS, 119991 Moscow, Russia

Abstract. Modified comparison method is developed. In this method detailed theoretical calculations of spectra of heavy elements is conducted varying one parameter (Te) in order to find the most complete matching of the structure of the theoretical and the experimental Mo spectra in laser-produced plasmas.

Development of new methods of diagnostics of high-temperature plasma of heavy elements, including Mo and W, is of great interest. Heavy elements are present in almost all controlled thermonuclear fusion devices: plasmas of powerful Z-pinches — the wire arrays material, plasmas of tokamaks - the walls and the divertor material. Spectroscopic diagnostics of such plasmas is a very difficult task due to the complex structure and huge number of lines in the spectra. To solve this problem, the comparison method [1,2] which is based on the comparison of the spectra of the investigated source with the spectra of well-diagnosed laser-produced plasmas, is developed further in this paper. The laser-produced plasmas diagnostics is performed using the spectra of [H] - and [He] - like ions by well-known and proven techniques.

Further comparison of the spectra of heavy elements in laser-produced plasmas with the spectra of the investigated source allows to evaluate the electron temperature T_e of the source. As an example, experimental W spectra of laser-produced plasmas at T_e =720±60 eV is shown in Fig. 1.

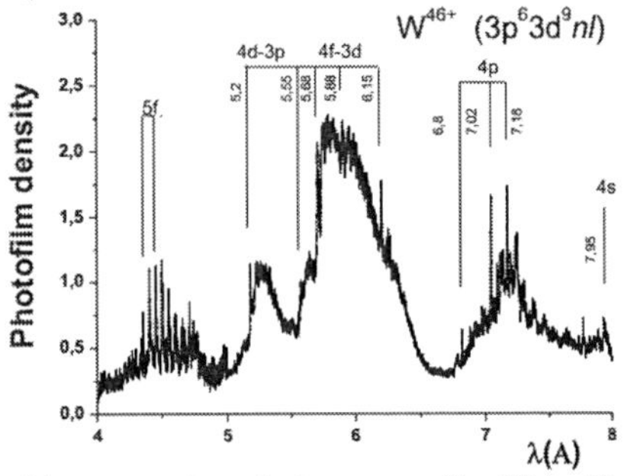

Fig. 1. Experimental W spectra of laser-produced plasmas at T_e=720±60 eV.

---

* Corresponding author: apshev51@gmail.com

This technique completely describes the structure of the X-ray spectra of W and Mo multiply charged ions, increases the accuracy of measurements and fully justifies the methods used. In particular, for the Mo laser-produced plasmas (n=3 - n=2 transitions in Mo^{31+} – Mo^{34+} ions) the electron temperatures determined experimentally by the comparison method (T_e = 685 eV) and calculated theoretically (T_e = 650 eV) coincided with a tolerance of less than (± 10) %. Fig.2 shows the comparison of experimental (Te = 685 eV) and theoretical (T_e = 650 eV) spectra of Mo.

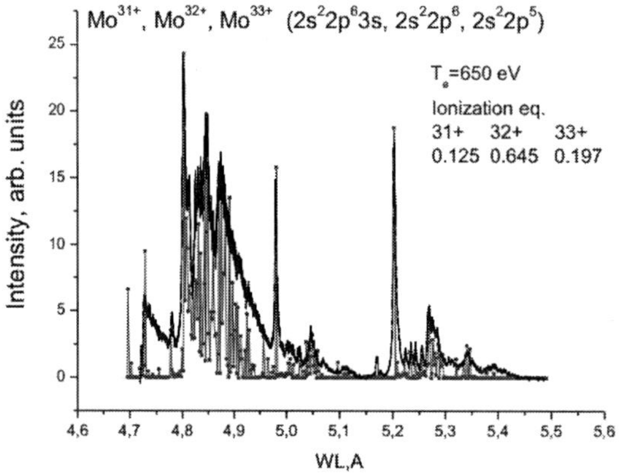

Fig. 2. Comparison of experimental (Te = 685 eV) and theoretical (T_e = 650 eV) spectra of Mo.

For the theoretical simulation of the experimental spectra the INDAHAUS code developed by the authors of the present paper (S.A. and I.T.) is used. The code is written in Python language which allows to run it in any platform (unix, windows, etc.). Ionization equilibrium and radiative-collisional characteristics being an input for the INDAHAUS code are calculated using GKU [3], ATOM [4] and FAC [5] codes. The INDAHAUS code is very flexible with respect to the number of ions included in the simulation and in the implementation of the various modes of behavior of plasma (coronal limit, the thermodynamic equilibrium, intermediate regimes).

After approbation of theoretical calculations on the experimental results for the high-temperature laser-produced plasmas, the theoretical spectra were calculated for higher temperatures: T_e> 1 keV, for which it is very difficult to carry out laboratory experiments. Thus, the modified method of comparison shows promising potential for the diagnostics of high-temperature plasma in fusion devices using the line x-ray spectra of multiply charged ions of heavy elements (W and Mo).

The support by Russian Foundation for Basic Research is acknowledged (15-02-04411).

References

1. A.P. Shevelko, Quantum Electronics **41**, 726 (2011)
2. A.P. Shevelko, D.E. Bliss et al., Plasma Physics Reports **34**, 944 (2008)
3. I.L. Beigman, Astrophysics and Space Physics Reviews **11**, 1 (2001)
4. V.P. Shevelko, L.A. Vainshtein, *Atomic Physics for Hot Plasmas* (IOP Publishing, Bristol, 1993)
5. V.P. Shevelko, L.A. Vainshtein, *Struktura i Kharakteristiki Ionov v Goryachei Plazme* (Nauka, Moscow, 1986) (in Russian)

Spectroscopy of systems of two identical atoms: effects of quantum interference

A.A. Makarov[1,2,*], and V.I. Yudson[1,3]

[1]Institute for Spectroscopy RAS, 108840 Moscow, Troitsk, Russia
[2]Moscow Institute of Physics and Technology, 141700 Dolgoprudny, Russia
[3]Russian Quantum Center, 143025 Skolkovo, Russia

Abstract. Several effects of quantum interference in spectroscopy of a system of two atoms are discussed. (i) In the system of spatially separated atoms in a one-dimensional (1D) geometry (a single-mode waveguide or photon crystal), a (meta)stable excited entangled state can be formed, its decay being very sensitive to the distance between the atoms and to perturbations which cause a difference between their resonance frequencies. (ii) In a system of closely located atoms in 3D space, the extreme sensitivity of absorption and fluorescence spectra to the direction of the applied magnetic field is demonstrated. These theoretical predictions can be useful for the quantum information processing and ultrasensitive measurements.

A system of two identical atoms is the simplest example that exhibits (i) the effect of entanglement which is of primary importance for the quantum information processing, and (ii) an extreme sensitivity of the absorption and emission spectra to the mutual position of the atoms. Only few experiments so far demonstrated these properties, but still far from their real potentiality. For instance, in Ref. [1], the fluorescence spectra of two trapped ions showed a weak dependence on the distance r between them, varied near $r \approx 2\lambda_{eg}$, where λ_{eg} was the wavelength of the used optical transition. On the other hand, theory predicts the two opposite limiting cases $r >> \lambda_{eg}$ and $r << \lambda_{eg}$ are more promising.

1D case of spatially separated atoms. If one of the atoms is initially excited its spontaneous decay leads to excitation of another atom, refection from it, and finally, if the distance between the atoms is equal to a whole number of the half-wavelengths of the optical transition, to the formation of a stable entangled state [2]. In the "ideal" case, this excited state is stable, and is formed with the 1/2 probability. In reality this state is meta-stable since any small deviation from the ideal condition leads to its decay. Such a perturbation can be caused by a deviation of the interatomic distance from the `ideal' one, and/or by different surroundings that lead to a detuning of the atom's transition frequencies [3]. So, the system can be useful both as a candidate for a long-lived entangled state, and as a precise tool for weak perturbation measurements.

3D case of closely spaced atoms in a magnetic field: Fluorescence spectra. A system of two identical atoms with the ground state $|g>$ and the excited state $|e>$ is an example exhibiting the effects of Dicke super- and subradiance. The two excited states of this system

* Corresponding author: amakarov@isan.troitsk.ru

are $|Q_1>=|e_1g_2>$ and $|Q_2>=|e_2g_1>$. The dipole-dipole interaction between the states $|Q_1>$ and $|Q_2>$ leads to their superpositions producing two entangled eigenstates, symmetric and antisymmetric. The symmetric state $(|Q_1>+|Q_2>)/\sqrt{2}$ is superradiant, decaying twice faster than the one-atom excited state whereas the antisymmetric state $(|Q_1>-|Q_2>)/\sqrt{2}$ is subradiant, decaying much slower (by the parameter $(r/\lambda_{eg})^2$ if the distance r between the atoms is smaller then the wavelength of the $|e>\to|g>$ transition. Now, the main idea is that a sufficiently strong magnetic field, directed at the magic $\alpha = \arccos(1/\sqrt{3}) \approx 54.7^0$ hinders the dipole-dipole interaction, so the fluorescent spectra become most simple. Moreover, they are very sensitive to a small deviation of the magnetic field from the magic angle. In this way, the vector separation between the atoms can be found. The second idea is the measurement of the gated fluorescence from the subradiant state only. Its relative yield gives an estimate for the distance between the atoms. So, the full nanoscopy of pairs of atoms can be achieved.

A procedure for switching from superradiance to subradiance._After selective excitation of the superradiant state and application of the strong magnetic field the following procedure can be applied. Its first act makes atoms non-identical in a sense of their transition frequencies. This can be achieved, e.g., by introducing a transverse gradient in the magnetic field. If this action is adiabatic then the superradiant state transfers to the product state, say $|e_1g_2>$. The second act is the instantaneous switching of the gradient direction that does not change the state but changes its energy. In the final act, the gradient is adiabatically turned off. Calculations show that the final state will be subradiant with the lifetime increased by $\sim(\lambda_{eg}/r)^2$ times. So, the actual goal of keeping quantum information for a long time with the use of subradiant states can be realized.

Narrow dip in the natural linewidth absorption profile. The first act above leads also to an interesting effect of the quantum interference in the absorption spectrum [4]. A typical example is shown in the figure 1 (where Γ is the radiation width and $k = 2\pi/\lambda_{eg}$). The dip's depth and the shape of the absorption profile depend on the distance between the atoms.

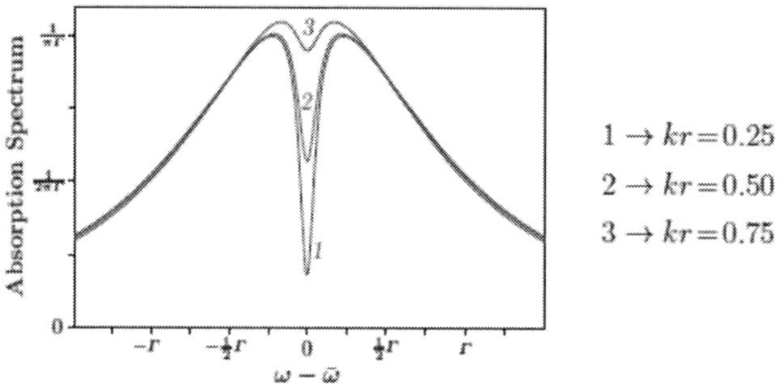

Fig. 1. Narrow dip in the natural linewidth absorption profile.

The work was supported by Russian Foundation for Basic Research (No. 15-02-05657).

References

1. R.G. DeVoe, R.G. Brewer, Phys. Rev. Lett. **76**, 2049 (1996)
2. A.A. Makarov, V.S. Letokhov, JETP **97**, 688 (2003)
3. E.S. Redchenko, V.I. Yudson, Phys. Rev. A **90**, 063829 (2014)
4. A.A. Makarov, Phys. Rev. A **92**, 053840 (2015)

Off Rowland spectrometer with shifted slit

P.S. Antsiferov[1,2,*], L.A. Dorokhin[1], P.V. Krainov[1,2]

[1]Institute for Spectroscopy RAS, 108840 Troitsk, Moscow, Russia
[2]Technological Institute for Superhard and Novel Carbon Materials, 108840, Troitsk, Moscow, Russia

Abstract. The article presents the scheme of grazing incidence spectrometer with the entrance slit shifted from the Rowland circle. This scheme allows to get the focusing surface normal to the line of site and to use the micro channel plate detector for the registration of the spectra. The results of the experimental tests are also given.

The detection of vacuum ultra violet (VUV) and soft X-ray radiation by means of micro channel plates (MCP) combines high sensitivity (quantum efficiency up to 15 %) and time resolution down to 1 ns, which is important for the plasma diagnostics applications. The usage of MCP detectors is optimal in the case of normal incidence of radiation. As it was shown still in [1], the shift of the entrance slit from the Rowland circle allows to change the angle of the radiation incidence on the surface of spectral focusing, see Fig.1.

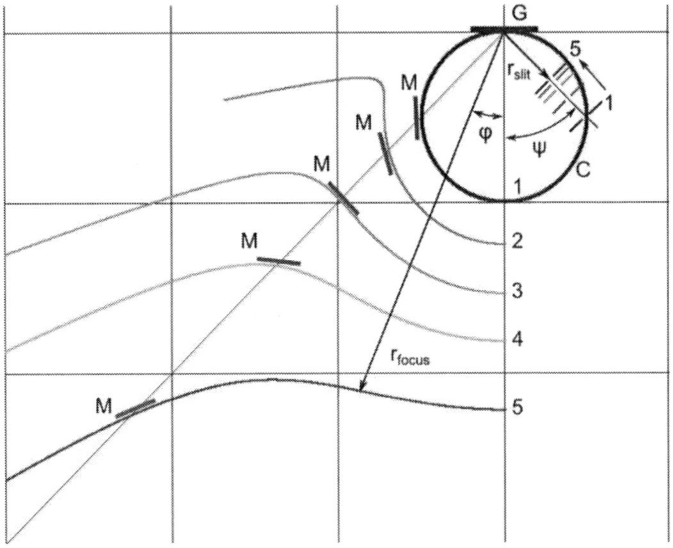

Fig. 1. G – diffraction grating, C – Rowland circle. The entrance slit positions 1 – 5 correspond to the focusing curves 1 – 5 according [1]. M – correspondent positions of the entrance surface of MCP.

* Corresponding author: Ants@isan.troitsk.ru

The particular scheme of the spectrometer for the wavelength range 100-300 Angstroms with MCP detector is given in Fig.2. The estimated spectral resolution $\lambda/\delta\lambda$ is in the range 300-700 [2]. The main parameters of the spectrometer: diffraction grating grooves density 600 gr/mm, diffraction grating radius of curvature 1m, the entrance slit width 10 μm, the grazing angle 4^0, the distance slit – grating 3.3 cm, the working size of the grating 6 mm, the distance MCP – grating 33.1 cm.

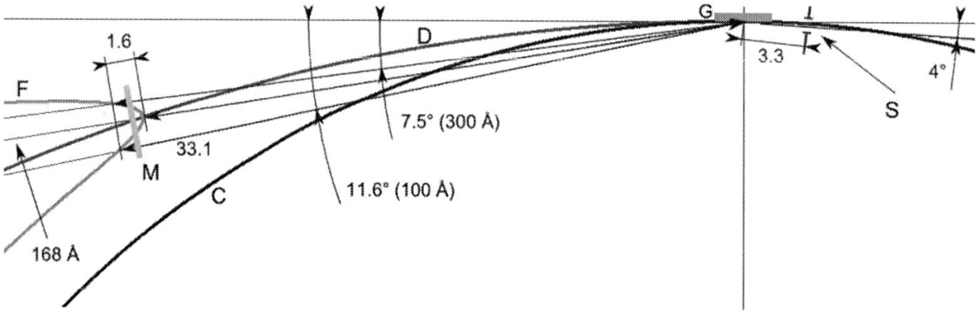

Fig. 2. The drawing of the proposed off-Rowland spectrometer with shifted slit. G – diffraction grating, R- Rowland circle, D – double Rowland circle, S – entrance slit, F – focusing curve, M – the entrance surface of the MCP.

An example of the VUV spectrum of multiply charged ions of Ar and the spectral resolution data are given in Fig.3. The deterioration of the experimental spectral resolution in comparison with calculated [2] can be explained by the influence of the apparatus function of MCP detector.

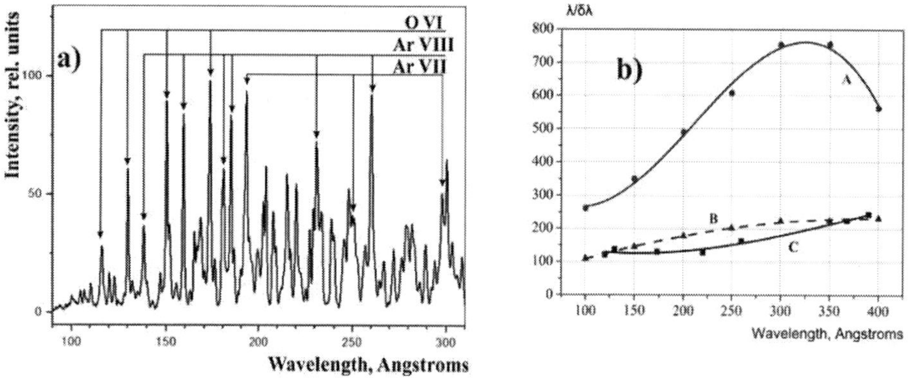

Fig. 3. a) the example of the spectrum, obtained with the help of off- Rowland spectrometer with shifted slit, b) A - the results of the numerical estimation [2] of the spectral resolution, B – the degradation of the spectral resolution by the influence of 100 μm apparatus function, C – the experimental results.

The work was supported by RFBR grant No 15-02-04411a.

References

1. W. Baily, Philisophical Magazine. Series 5 **15**, 183 (1883)
2. P.S. Antsiferov, L.A. Dorokhin, P.V. Krainov, Review of Scientific Instruments **87**, 053106 (2016)

Study of optical and luminescent properties of nanocrystals $NaYF_4:Tm^{3+}$, Yb^{3+} in the UV range in the application of integrated optics

I.M. Asharchuk[1,*], S.I. Molchanova[1], V.V. Rocheva[1], M.S. Baranov[2], M.E. Sarycheva[3], and K.V. Khaydukov[1]

[1]FSRC "Crystallography and Photonics" RAS, "Institute of Photonic Technology" RAS, 142092 Moscow, Troitsk, Russia
[2]Volgograd State University, 400062 Volgograd, Russia
[3]Moscow Technological University, 119571 Moscow, Russia

Abstract. Studied the photoluminescence properties of synthesized nanocrystals doped with rare-earth ions NaYF4:Tm^{3+}, Yb^{3+}, measured luminescence spectra and absorption in the visible and near infrared regions of 300-1000 nm. Were measured the energy of phonons these nanocrystals, the average phonon energy was 332cm^{-1}. Made optical waveguide impregnated with nanoparticles $NaYF_4$: Yb^{3+}, Tm^{3+} as the prospect of a compact source of radiation in the visible and UV range.

Recently, luminescent properties of materials doped with rare earth (RE) ions are attracted great attention due to their special characteristics: rich structure of the energy levels, low phonon energy, narrow lines of photoluminescence (PL) and a prolonged lifetime of the excited states. Such materials are nanoparticles, which are crystalline matrix of β-$NaYF_4$ a hexagonal structure, doped with rare-earth ions Tm^{3+}, Yb^{3+}. Yb^{3+} are used as donors transmit the absorbed energy-acceptor ions (Er^{3+}, Tm^{3+}) to increase efficiency upconversion. Ytterbium ion has broad absorption band in the range 970-980 nm [1].

Synthesized monodisperse upconversion nanoparticles with a core-shell structure with a size of 70 to 80 nm by the method described in [2]. The core - $NaYF_4$ simultaneously doped Yb^{3+}, Tm^{3+}, shell - $NaYF_4$.

To study the optical and photoluminescence properties of stable colloid was prepared using 2.966 g trihlometana and adding 0.1% by weight of upconversion nanoparticles. To nanoparticles were uniformly distributed over the volume of the solvent mixture by ultrasonic bath for 20 minutes.

Nanocrystals PL spectrum measurements were carried out using a spectrofluorometer Fluorolog3 manufactured by Horiba Jobin Yvon. The luminescence spectrum, taken at room temperature, is shown in Fig. 1a. It is evident that nanocrystals exhibit intense photoluminescence lines in the visible and near-IR spectral range.

* Corresponding author: ilya-asharchuk@yandex.ru

DOI: 10.1051/epjconf/201713203002

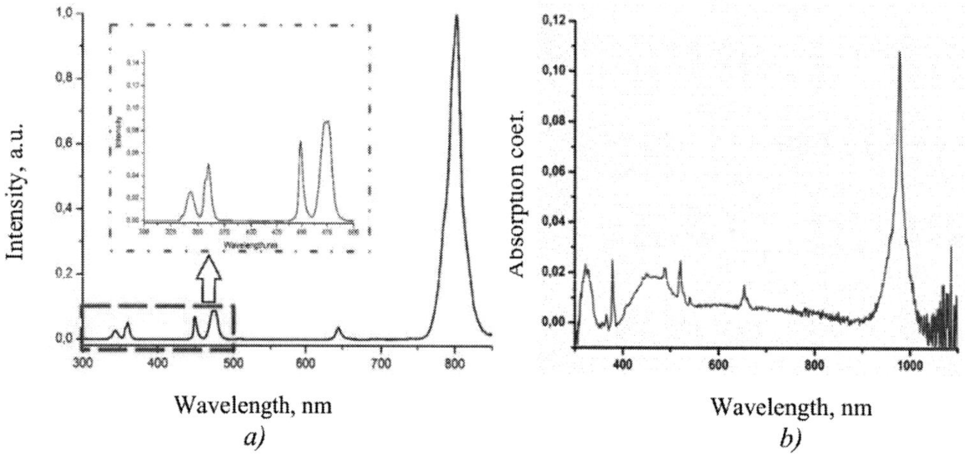

Fig. 1. The photoluminescence spectra of (a) and absorption (b) nanoparticles $NaYF_4$: Yb^{3+}, Tm^{3+} @ $NaYF_4$

To determine the phonon spectrum used Raman spectrometer LABRam HR800 Horiba JY, a test signal is a light He-Ne laser with a wavelength of 632.8 nm, because this wavelength does not fall in the absorption band of the material (1b). In this crystalline matrix, there are several phonon frequencies from 291 cm^{-1} up to 413 cm^{-1}.

For the manufacture of the waveguide used photocurable polymeric material SU-8 production MicroChem, in which nanoparticles are impregnated, by adding 0.1% by weight of the nanocrystals followed ultrasonic mixture. Light-guiding core formed by contact UV photolithography on a silicon substrate. waveguide length was 3 cm from the light-guiding core size 8x8 mm. The results obtained in the current work demonstrates the possibility of using nanoscale crystals synthesized in integrated optics to create the compact sources of radiation in the ultraviolet and visible regions of the spectrum.

Research supported byRFBR grant 16-34-00934-a in the synthesis of the nanoparticles and their photoluminescence properties, RFBR grant 14-07-00759A part in the formation of polymer waveguides impregnated with nanoparticles.

References

1. L.D. DeLoach, S.A. Payne, L.L. Chase et al., J. Quantum Electronics **29**, 1179 (1993)
2. E.A. Grebenik, A. Nadort, A.N. Generalova et al., J. Biomed. Opt. **18**, 076004 (2013)

Kinematic anharmonicity of internal rotation of molecules

V.A. Bataev*, V.I. Pupyshev, and I.A. Godunov

Lomonosov Moscow State University, Faculty of Chemistry, 119991 Moscow, Russia,

Abstract. The methods of analysis the strongly coupled vibrations are proposed for a number of molecules of aromatic and heterocyclic carbonyl (and some others) compounds. The qualitative principles are formulated for molecular systems with a significant kinematic anharmonicity.

In our report, we demonstrate methods of analysis of the strongly coupled molecular vibrations and formulate quality criteria to search of molecular systems with significant kinematic anharmonicity. As illustrations the molecules of aromatic and heterocyclic carbonyl (and some other) compounds are chosen.

The special attention was focuses on interactions of the rotation of the carbonyl group with other non-planar vibrations. It is interesting, that for the systems considered the form of the potential energy surface (PES) do not demonstrate any essential interactions of vibrational modes. The observed interaction is primarily due to the kinematic anharmonicity, which is manifested in a relatively large amplitude of changes of the off-diagonal elements of the kinematic matrix in some regions of PES.

The detailed analysis of internal rotation was performed for the benzaldehyde (C_6H_5CHO) and furfural (C_5H_4OCHO) molecules in the ground electronic states by methods of quantum chemistry. For the systems under investigation, this is a complex task, that requires the methods more accurate than the standard one-dimensional (1D) approximation. Really, one may find, for example, essential differences (especially for benzaldehyde molecule) between the experimental and calculated values of the potential barriers to internal rotation. For the molecules mentioned it was found the non-trivial situation of significant kinematic (but not dynamic) interaction between torsional vibration and out-of-plane CHO deformation [1,2]. Taking into account this kinematic interaction one may find the good agreement between experimental and calculated energies of torsional and out-of-plane CHO deformation energy levels within the two-dimensional (2D) approximation for the vibrational problem, while disagreement in energies for traditional 1D approximations are essential. It should be noted, that in the case of benzaldehyde molecule such agreement is reached, provided correct errors in assignment of torsional transitions in the experimental vibrational spectrum [1]. In particular, the results presented explain the disagreement in high level nonempirical and simplified 1D-estimates of barriers to internal rotation.

* Corresponding author: v.a.bataev@gmail.com

DOI: 10.1051/epjconf/201713203003

The features of the internal rotation were also studied for other molecular systems containing benzene or various heterocyclic rings and the functional group (-CHO, -CDO, -CFO et al.). It is shown that in some of the systems examined the relationship of rotation of the functional groups associated with the ring, and non-planar vibrations is predominantly kinematic, and is determined mainly by geometric parameters of this group and the corresponding distribution of point masses. The dependence of the kinematic effects on the mass distribution within the benzene or heterocyclic ring is much weaker. In contrast to the benzaldehyde molecule, the 1D potential function of internal rotation for related molecule of benzoyl fluoride (C_6H_5CFO) calculated as the minimal energy path is quite reliable; as a result, the 1D and 2D energies of the torsional energy levels are practically the same. This example and the results of testing of the similar molecules C_6H_5–CXY demonstrate clearly that one may expect the significant kinematic interaction between torsional vibration and out-of-plane CXY deformation for the systems where the masses of atoms X and Y differ essentially, as it is in the benzaldehyde molecule, but not in the benzoyl fluoride molecule [1].

The support from Russian Foundation for Basic Research is acknowledged (16-03-00794a).

References

1. I.A. Godunov, V.A.Bataev, A.V. Abramenkov, V.I. Pupyshev, J. Phys. Chem. A **118**, 10159 (2014)
2. V.A.Bataev, A.V. Abramenkov, V.I. Pupyshev, I.A. Godunov, Spectrochim. Acta Part A **161**, 155 (2016)

EPJ Web of Conferences **132**, 03004 (2017) DOI: 10.1051/epjconf/201713203004
SPECTROSCOPY.SU 2016

Deformation splittings in the spectra of $LaAlO_3$: Ho^{3+}, Pr^{3+}, Tm^{3+} single crystals

K.N. Boldyrev[1,*], P. Dereń[2], and M.N. Popova[1]

[1]Institute for Spectroscopy RAS, 108840 Troitsk, Moscow, Russia
[2]Institute of Low Temperature and Structure Research, Polish Academy of Sciences, 50-950 Wroclaw, Poland

Abstract. The work is devoted to a high-resolution spectroscopy study of the perovskite-type single crystals of $LaAlO_3$ doped with non-Kramers rare-earth ions of praseodymium, holmium and thulium. Doublets of a specific shape were observed in the spectra, which we attribute to the presence of random deformations in the crystals.

It has been shown earlier that for dielectric crystals doped with rare-earth (RE) ions random deformations of the crystal lattice may result in a specific shape (with a narrow dip in the center) of spectral lines corresponding to transitions between crystal-field (CF) singlets (or Kramers doublets) and non-Kramers doublets [1-3]. In particular, a well resolved splitting of RE spectral lines caused by random lattice strains was observed in the spectra of tetragonal centers in $LiYF_4:Tm^{3+}$ [1] and $CaWO_4:Ho^{3+}$ [3] crystals and of cubic centers in hexafluoroelpasolite crystals doped with Sm, Er, Tm, or Yb [2]. It is of interest to observe such effect in the spectra of trigonal centers, in which non-Kramers ions also possess doubly degenerate CF levels.

We have chosen a $LaAlO_3$ crystal that belongs to the family of perovskites, possesses a pseudo cubic structure, and has a promising application potential. RE^{3+} ions in this compound substitute for La^{3+} which occupies a single crystallographic position with the D_3 point symmetry. There are Γ_1 and Γ_2 singlet and Γ_3 doublet CF levels for a non Kramers RE^{3+} ion residing in the D_3 symmetry position. Optical properties of $LaAlO_3$ doped with Pr^{3+}, Ho^{3+}, and Tm^{3+} ions were studied in Refs. [4-7] under a medium spectral resolution. In this work, we have measured high-resolution (0.02 cm^{-1}) broad-band (2000-20000 cm^{-1}) temperature-dependent transmission spectra of $LaAlO_3$ single crystals doped with non Kramers RE ions Ho^{3+}, Pr^{3+}, or Tm^{3+}. A splitting of several spectral lines corresponding to the Γ_1, $\Gamma_2 \rightarrow \Gamma_3$ transitions was observed (see, e.g., Fig. 1). We attribute this splitting to a combined action of the magnetic hyperfine interaction and random deformations of the crystal lattice.

The theory developed in Refs. [1-3] enabled to successfully model the line shapes and gave a tool for getting a quantitative estimate for the concentration of intrinsic lattice defects, by comparing experimental and calculated spectral shapes. To perform such calculations, the CF and electron-deformation interaction parameters are necessary. These

* Corresponding author: kn.boldyrev@gmail.com

parameters can be obtained from the CF calculations based on the experimentally found scheme of the CF levels and from the data of piezospectroscopic measurements. In this work, by analyzing high-resolution temperature-dependent transmission spectra of $LaAlO_3:Pr^{3+}$(0.1 at. %) single crystals we have determined energies of the CF levels of Pr^{3+} in the energy range up to 20000 cm^{-1}.

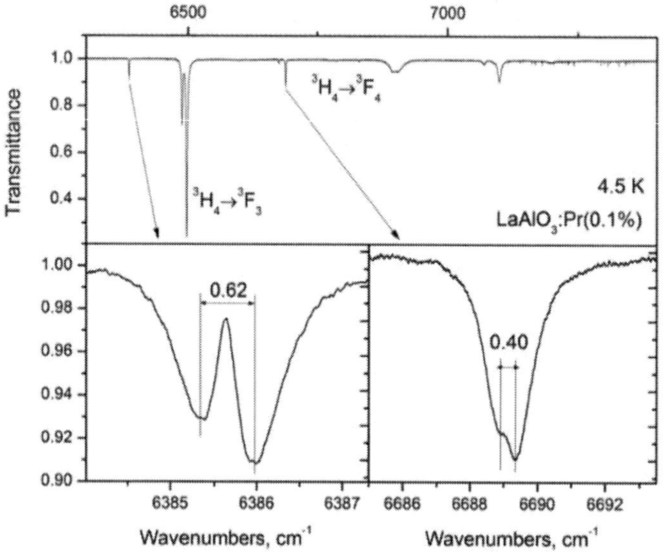

Fig. 1. A part of the Pr^{3+} spectrum in a $LaAlO_3:Pr^{3+}$(0.1 at. %) single crystal at the temperature 4.5 K. The lower part of the figure shows a doublet structure of some lines.

This dataset forms a basis for the CF calculations, a further modeling of deformation splittings observed in the spectra of trigonal RE centers, and, in this way, for development of a method to estimate the quality of perovskite crystals, which are known for a vast variety of applications.

References

1. S.A. Klimin, D.S. Pytalev, M.N. Popova, B.Z. Malkin, M.V. Vanyunin, S.L. Korableva, Phys. Rev. B **81**, 045113 (2010)
2. B.Z. Malkin, D.S. Pytalev, M.N. Popova, E.I. Baibekov, M.L. Falin, K.I. Gerasimov, N.M. Khaidukov, Phys. Rev. B **86**, 124406 (2012)
3. G.S. Shakurov, E.P. Chukalina, M.N. Popova, B.Z. Malkin, A.M. Tkachuk. Phys. Chem. Chem. Phys. **16**, 24727 (2014)
4. P. Dereń, J. Lumin. **122–123**, 40 (2007)
5. M. Malinowski, M. Kaczkan, S. Turczyński, D. Pawlak, Opt. Mater. **33**, 1004 (2011)
6. P. Dereń, J.-C. Krupa, J. Alloys Comp. **380**, 362 (2004)
7. A. Gocalińska, P. Dereń, P. Głuchowski et.al., Opt. Mater. **30**, 680 (2008)

Analysis of vibronic interactions in the molecules of cross-conjugated ketones

V.V. Kompaneez*, and I.A. Vasilieva

Moscow State Pedagogical University, 119991 Moscow, Russia

Abstract. We have done quantitative analysis of vibronic parameters of two cross-conjugated δ-dimethylaminoketones. The research shows the influence of C-N and C=O bonds in the rings, and the radicals with nitro compounds on the vibronic parameters of characteristic bands, which describe the state (vibrations, types of deformation under excitation) of the phenyl ring and the polyene bridge. Results described impact of the substituent's nature on the parameters of intra- and intermolecular interactions presents for the studied compounds.

We have done analyses, which interpreting major characteristic vibrational bands cross-substituted ketones. It shows that ketones like the diphenylpolyenes, one can observe the appearance of the lines in the spectra which corresponding to the change of such specific elements molecules as polyene bridge or aromatic ring.

It is known that frequency bands of compounds with the carbonyl group in six-bonds ring are the same as in the compounds with the open chain (when C=O group is located between two methylene groups) [1], however, it known nothing about the intensity of these bands. Therefore, it was interest to produce the comparative analysis of the spectral properties of ketones and diphenylpolyenes at the same time, and quantification of the contribution of intramolecular interactions in the organization of the fine-structure spectra of these molecules. In order to realize such kind if investigations the different methods of selective laser spectroscopy can be used [2, 3], in particular vibrational analysis of conjugated absorption and luminescence spectra [4-7].

In this study, the basic osciliation for ketones with general formula $R_{1,2}$=Ph–CO–R_1 (where $R_1 \rightarrow (CH{=}CH)_2{-}N(CH_3)_2$, $R_2 \rightarrow CH_3{-}NC_4H_3{-}CH$, Ph→$(HOC_6H_6)$) were correlated with the interpretation of similar value and substituent frequencies 1,4-diphenylbutadiene with the substituent on the ring $N(CH_3)_2$ and NO_2, respectively (after - DFB3), which spectral properties have been studied previously [8-10]. We shown how band intensities ketones characterizing polyene bridge and the phenyl ring are changes. We selected the first molecule of the ketone with the same substituents as locate in DFB3, left polyene chain in the second molecule of the ketone folds into a five-membered ring (R_2).

It is shown that the parameters of vibronic spectra of ketone molecules exhibit a high degree of tolerance in the molecules sequence which having the same set of structural elements. It greatly expands the possibilities for use of the fragmented approach to describe the fundamental bands of molecules [6, 7, 11-13].

Participation ketone molecules in the formation of hydrogen bonds significantly reduces the frequency of the C=O (Tab. 1, line 2), conjugate with multiple bonds and aromatic rings

* Corresponding author: kompaneez@gmail.com

significantly affect by the position and intensity of the band in the range of 1500-1760 cm^{-1}. The implantation of substituents and representation polyene chain as cycle form with a heteroatom, which containing two multiple bonds, decreases the dipole moment of the molecule. It lead to reduces the intermolecular interaction parameters. When we considering the bands with weak manifestation in the spectra of some luminophors, where C-H and C=C bonds fluctuations occur, we confirmed that the presence of substitutes and the aromatic rings can be used to enhance the intensity of these bands (see Tab.1, line 1).

Table 1. The charts of parameters intramolecular (FC) and intermolecular (HT) interactions.

<table>
<tr><th>Interpretation</th><th>Vibration of parameters FC</th><th>Vibration of parameters HT</th></tr>
<tr><td>Planar and plane deformation vibrations of C-H bonds substituted aromatic components at the double bond.</td><td>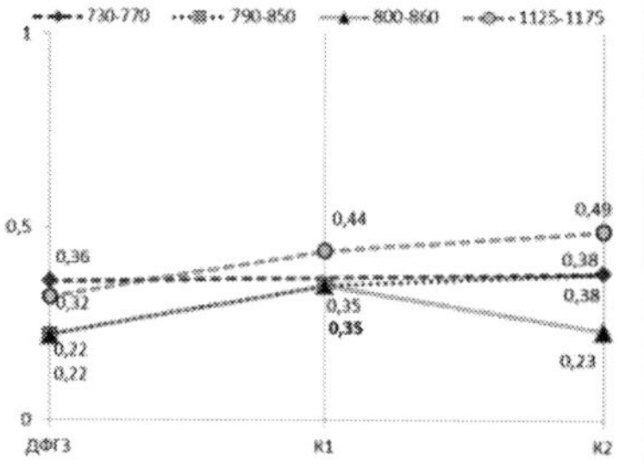</td><td>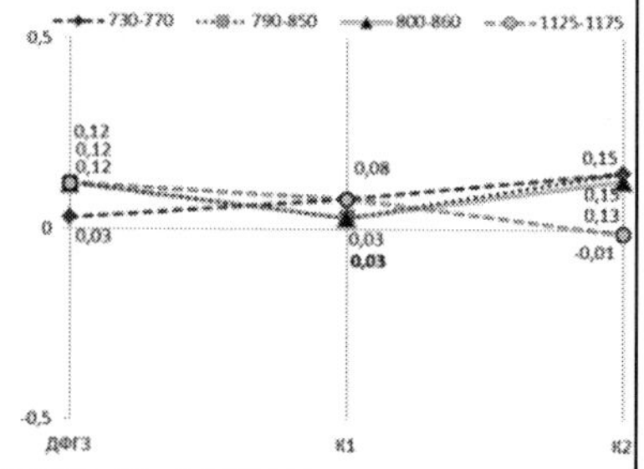</td></tr>
<tr><td>Deformation vibrations of N-H, a manifestation of C-N bond in the cycle, the manifestation of aryl nitro compounds R-NO_2. Stretching vibrations of the C = C, coupled with the C = O.</td><td>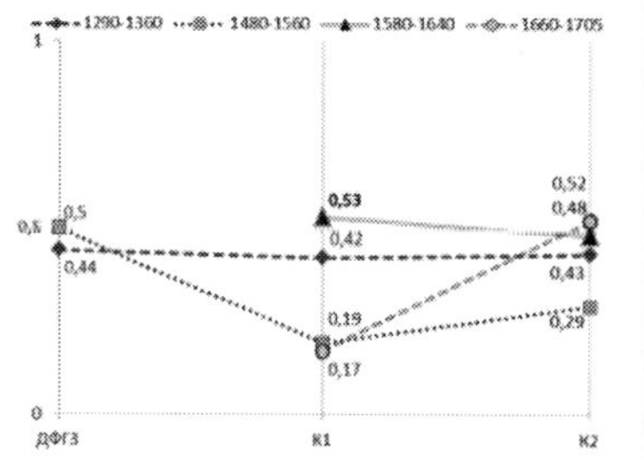</td><td>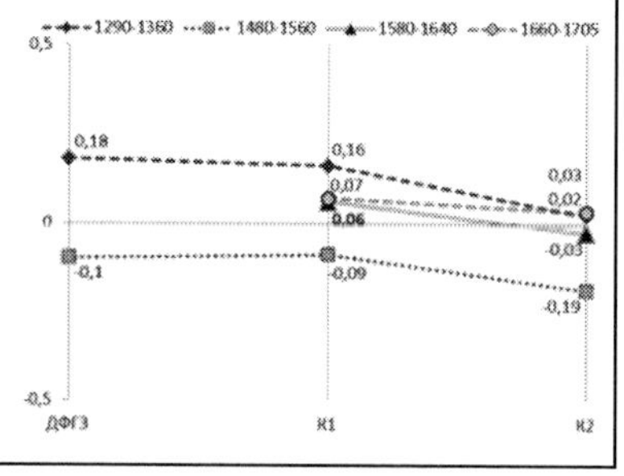</td></tr>
</table>

This work was financially supported by the Russian Foundation for Basic Research, project no. 15-32-21100-mol_a_ved.

References

1. L.J. Bellamy, *The Infrared Spectra of Complex Molecules* (Springer, Berlin, 1975)
2. M. Orrit, J. Bernard, R.I. Personov, J. Phys. Chem. **97**, 10256 (1993)
3. A.V. Naumov, Physics Uspekhi **56**, 605 (2013)
4. A.V. Naumov, N.L. Naumova, I.S. Osad'ko, Opt. Spectrosc. **91**, 750 (2001)
5. N.L. Naumova et al., Opt. Spectrosc. **92**, 426 (2002)
6. N.L. Naumova et al., Opt. Spectrosc. **98**, 535 (2005)
7. N.L. Naumova, I.A. Vasil'eva, I.S. Osad'ko, A.V. Naumov, J. Lumin. **111**, 37 (2005)
8. V.V. Kompaneets, I.A. Vasilieva, Bull. RAS: Physics **80**, 848 (2016)
9. I.A. Vasil'eva et al., J. of Optical Technology **78**, 286 (2011)
10. I.A. Vasilieva, V.V. Kompaneets, Opt. Spectrosc. **114**, 60 (2013)
11. I.A. Vasilieva, V.V. Kompaneets, Opt. Spectrosc. **115**, 32 (2013)
12. V.V. Kompaneets, I.A. Vasilieva, Opt. Spectrosc. (2016) (to be published)
13. I.A. Vasil'eva, N.A. Voitova, R.N. Nurmukhametov, Opt. Spectrosc. **112**, 488 (2012)

Near-field polarization distribution of Si nanoparticles near substrate

S.A. Reshetov[1], Yu.V. Vladimirova[1,*], L.P. Gevorkian[2], and V.N. Zadkov[1,2,3]

[1]International Laser Center and Faculty of Physics, Lomonosov Moscow State University, 119991 Moscow, Russia
[2]Kolmogorov School, Lomonosov Moscow State University, 119991 Moscow, Russia
[3]Institute for Spectroscopy RAS, 108840 Troitsk, Moscow, Russia

Abstract. Structure of the near-field intensity and polarization distributions, the latter is described with the generalized 3D Stokes parameters, of a spherical Si subwavelength nanoparticle in a non-magnetic and non-absorbing media near a dielectric substrate has been studied in detail with the help of the Mie theory and an extension of the Weyl's method for the calculation of the reflection of dipole radiation by a flat surface. It is shown that for the nanoparticle near the substrate the interference effects due to the scattering by the nanoparticle and interaction with the substrate play an essential role. We also demonstrate how these effects depend on the dielectric properties of the nanoparticle, its size, distance to the substrate as well as on the polarization, wavelength and incident angle of the external light field.

Control of the near-field in the proximity of nanostructures is a key to shaping the spatial intensity of light and its polarization distribution at the nanoscale [1]. Near-field being formed by the interference of the incident electromagnetic field with the local field of the nanoparticle strongly depends both on polarization of the incident field and properties of the nanoparticles (their shape and materials). The orientation of the near-field polarization is a key quantity in many theoretical studies on nanooptics, nanophotonic devices and optical sciences in general. Recently, we studied theoretically how the near-field polarization distribution of a plasmonic prolate nanospheroid interacting with a plane electromagnetic wave depends on the polarization and frequency of the incident electromagnetic field [2, 3].

Along with plasmonic nanoparticles, the nanoparticles made of high refractive dielectric or semiconductor materials have recently received considerable attention in the nanophotonics for their ability to control and manipulate the light in the near-field [4, 5]. They allow direct engineering a magnetic field response at optical frequencies in addition to the electric field response in plasmonic structures. As a fundamental building block, the dielectric spherical nanoparticle presents both strong magnetic dipole and electric dipolar responses corresponding to the basic Mie resonances. In many real applications the nanoparticles are often located near a substrate and the influence of the substrate may be

* Corresponding author: yu.vladimirova@physics.msu.ru

crucial. In our work, we have studied a structure of the near-field of a spherical Si nanoparticle in a non-magnetic and non-absorbing media near a dielectric substrate with the help of the Mie theory for scattering by a sphere in a homogeneous medium and an extension of the Weyl's method for calculation of the reflection of dipole radiation by a flat surface [5, 6]. The intensity and polarization distributions have been obtained by this method.

We consider two types of substrates with the different absorption coefficients: Im n≈1.5 (carbon) and Im n≈0 (silicon dioxide) (Fig. 1). Fig. 2 shows the distribution of the near-field polarization under the circular polarized incident field for Si nanoparticles without a substrate (left plot), near a quartz substrate (middle plot) and near a graphite substrate (right plot). Significant differences for the distributions near a substrate are due to the Fresnel reflection coefficients, which result in a phase shift in each plane wave in the expansion of the field.

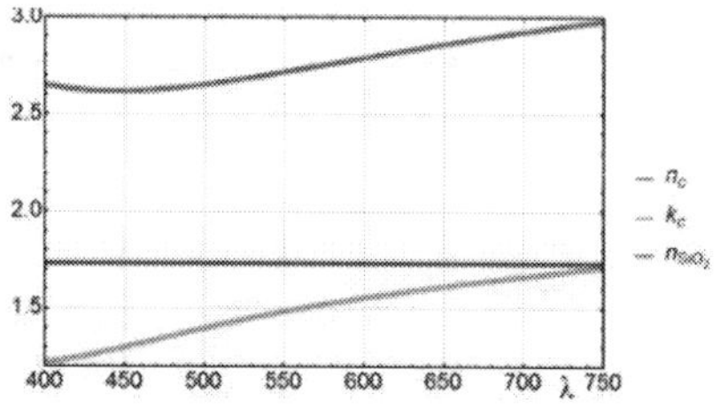

Fig 1. Optical constants of carbon and SiO_2

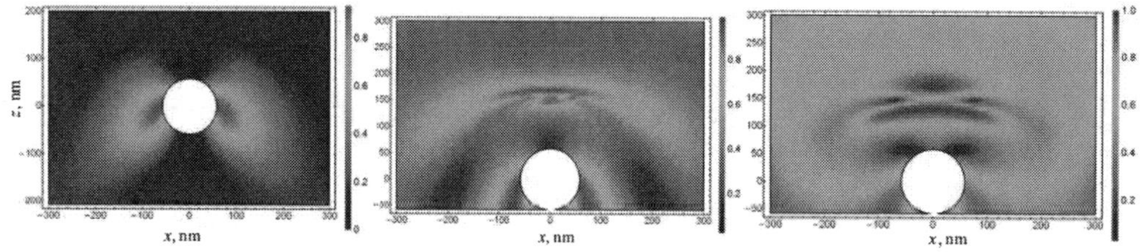

Fig 2. The distribution of the degree of polarization ($P=(I_{max}-I_{min})/(I_{max}+I_{min})$, where I_{max} and I_{min} are the maximum and minimum intensities of the near-field in a specitif point of space around the nnaoparticle) of the near-field under the circular polarized incident field for Si nanoparticles without a substrate (left plot), quartz substrate (middle plot) and graphite substrate (right plot).

In terms of the generalized 3D Stokes parameters, we analyzed the influence of the substrate on the structure of the near-field versus such parameters of the system as the radius and distance of the particles to the surface, polarization, wavelength, the angle of the incident wave, and the dielectric constant of all elements of the system. It is shown that for the nanoparticle near the substrate the interference effects due to the scattering by the nanoparticle and interaction with the substrate play an essential role leading specifically to much more complicated polarization distribution of the near-field by contrast with the case of a nanoparticle in free space filled with a homogeneous medium.

The authors would like to acknowledge financial support from the Russian Foundation for Basic Research under the grant No. 16-02-00816.

References

1. B. Hecht, L. Novotny, *Principles of Nano-Optics* (Cambridge University, 2006)
2. E.D. Chubchev, Yu.V. Vladimirova, V.N. Zadkov, Opt. Express **22**, 20432 (2014)
3. E.D. Chubchev, Yu.V.Vladimirova, V.N. Zadkov, Laser Phys. Lett. **12**, 015302 (2015).
4. D. Permyakov et al., Appl. Phys. Lett. **106**, 171110 (2015)
5. A.E. Miroshnichenko, A.B. Evlyukhin, Yu.S. Kivshar, B.N. Chichkov, ACS Photonics **2**, 1423 (2015)

EPJ Web of Conferences **132**, 03007 (2017) DOI: 10.1051/epjconf/201713203007
SPECTROSCOPY.SU 2016

A many-particle quantum-kinetic formalism for describing properties of light emitters in frozen dielectrics

M.G. Gladush[1,2,3,*]

[1]Institute for Spectroscopy RAS, 108840 Troitsk, Moscow, Russia
[2]Moscow State Pedagogical University, 119991 Moscow, Russia
[3]Moscow Institute of Physics and Technology, 1417009 Dolgoprudny, Russia

Abstract. A many particle quantum-kinetic formalism is suggested to derive the Maxwell-Bloch-type equations which describe the interaction of quantum emitters with light in a frozen dielectric. It is shown that the quantum-kinetic formalism can meet the concept of local variations of dielectric properties and their influence on the emitter. The definitions of the local response and the effective refractive index in macroscopically homogeneous media are discussed.

We suggest a rigorous methodological approach that is essential in the theory of light emission by either a single quantum system or a light emitting ensemble when dynamics of the excited states and radiative properties are essentially dependent on its macroscopic dielectric host and the nearby surroundings. Most experimental data suggest that the properties of light from a single emitter are determined by the environment at the micro-, meso-, and macroscales. The corresponding analytical theories, however, remain either model-dependent or limited to describing specific situations and effects. Many studies found in the literature have been aimed at determining the change in the rate of the radiative decay of an excited single particle in a model environment as compared to its fluorescence in a material vacuum. These approaches are conventionally divided into "microscopic" and "macroscopic" models [1]. The common points in the models are the attempts to evaluate modifications of intrinsic properties of the emitting particle, the structure of the local driving field and the change in the local density of photonic states.

This research is focused on the properties of quantum light emitters embedded in a host material with likely local inhomogeneities in the presence of an external cw-laser beam. The mathematical tools used in laser physics, nonlinear and quantum optics, and other research related to interaction of radiation with the matter is largely based on an analysis of Maxwell-Bloch (MB) equations. To date, there has been a limited number of papers suggesting MB equations derived in a consistent manner and which include the effective values of all parameters: the Rabi frequency, transition frequency shifts and the rates of relaxation/excitation mechanisms (including spontaneous emission and dipole-dipole interactions) [1, 2]. In this work, we have made a joint between a sophisticated system of

* Corresponding author: mglad@isan.troitsk.ru

MB-type equations and observations of the excitations lifetime dispersion of single molecules (SM) in frozen matrices [3-5]. The lifetimes of the SM excited state can be measured using the data from SM zero-phonon line (ZPL) spectroscopy [6], at the conditions when the contribution from electron-phonon and electron-tunneling couplings are negligible. These conditions must be justified by providing extra accuracy in measuring ZPL power saturation and temperature dependences for each SM [6-9]. The excitation lifetime is a parameter in MB equations and must agree with the observations.

The MB-type system was derived for emitters in an ensemble of motionless particles embedded in a dielectric medium, which is transparent for the incident light. It includes the effective rates of individual and collective radiative damping, the Rabi frequencies, and frequency shifts of the transitions caused by the presence of the dielectric host and other quantum objects. These MB parameters were found to be functions of the real and imaginary parts of the host's permittivity. The excitation lifetimes of the emitters were shown to be functions of the effective refractive index and the extinction coefficient of the host around the emitter and agree with the experimental data.

Additionally, this work suggests a discussion of the results of the nanoscale mapping of the effective refractive index values in frozen dye-doped solutions performed by single-molecule spectromicroscopy [10,11]. It provides an estimation of the minimum volume of the host material which one may attribute to the effective refractive index.

This work was supported by RFBR (proj. nos. 14-29-07270, 16-02-01174).

References

1. D.V. Kuznetsov, V.K. Roerich, M.G. Gladush, J. Exp. Theor. Phys. **113**, 647 (2011)
2. D.V. Kuznetsov, V.K. Roerich, M.G. Gladush, Theor. Math. Phys. **168**, 1078 (2011)
3. E.A. Donley, V. Burzomato, U.P. Wild, T. Plakhotnik, J. Lumin. **83-84**, 255 (1999)
4. E.A. Donley, S. Bonsma, V. Palm, V. Burzomato, U.P. Wild, T. Plakhotnik, J. Lumin. **87-89**, 109 (2000)
5. E.A. Donley, T. Plakhotnik, J. Chem. Phys. **114**, 9993 (2001)
6. A.V. Naumov, Physics Uspekhi **56**, 605 (2013).
7. A.V. Naumov, Y.G. Vainer, M. Bauer, L. Kador, Phys. Stat. Sol. B **241**, 3487 (2004)
8. A.V. Naumov, Y.G. Vainer, L. Kador, Phys. Rev. B **79**, 132201 (2009)
9. K.R. Karimullin, A.V. Naumov, J. Lumin. **152**, 15 (2014)
10. A.V. Naumov, I.Y. Eremchev, A.A. Gorshelev, Eur. Phys. J. D **68**, 348 (2014)
11. T.A. Anikushina, M.G. Gladush, A.A. Gorshelev, A.V. Naumov, Faraday Discussions **184**, 263 (2015)

New methods of statistical processing of single-molecule spectromicroscopy data for mapping of local fields in the layer of a host matrix

A.V. Golovanova[1,2,*], T.A. Anikushina[1,2], A.A. Gorshelev[1], O.N. Korotaev[2], and A.V. Naumov[1,2]

[1]Institute for Spectroscopy RAS, 108840 Troitsk, Moscow, Russia
[2]Moscow State Pedagogical University, 119991 Moscow, Russia

Abstract. We describe the specific algorithms for statistical processing of the data on fluorescence excitation spectra and images for large number of single dye molecules in thin solid film. Analysis of the data allowed mapping of the local fields, electron-phonon coupling parameters, and the effective refractive index in a frozen solid film. An estimation of the volume attributed to the effective local values is given.

Laser fluorescence spectromicroscopy of single probe molecules (SMSM) in transparent solids is a widely recognized area of science [1, 2]. The SMSM demonstrates its best performance at cryo-temperatures, when zero-phonon lines (ZPL) of light emitters can be detected with the use of a narrowband laser [3, 4]. The point is single molecule (SM) ZPL is very sensitive to any kind of local fields and their fluctuations [3-8]. In particular the homogeneous SM ZPL width depends on the temperature as following:

$$\Gamma_{ZPL}(T) = \Gamma_0 + \Delta\Gamma_{e-tunn}(T, t_m) + \Delta\Gamma_{e-phon}(T), \tag{1}$$

where Γ_0 is the local-field depended life-time limited ZPL linewidth, $\Delta\Gamma_{e\text{-}tunn}(T, t_m)$ is the contribution to the width of ZPL, which is determined by the interaction of SM with tunneling elementary excitations in the matrix, $\Delta\Gamma_{e\text{-}phon}(T)$ is the broadening of the ZPL caused by quadratic electron-phonon coupling. Here

$$\Gamma_0 = \Gamma_0(n) = \frac{1}{2\pi T_1} = \frac{n \cdot l(n)}{2\pi \cdot \tau_0}, \quad \Delta\Gamma_{e-phon}(T) = B\frac{\exp(-h\nu/kT)}{[1-\exp(-h\nu/kT)]^2}, \tag{2}$$

where n is effective refractive index of the matrix, τ_0 is vacuum natural excited state lifetime, ν is effective frequency of the quasi-localized low-frequency vibrational mode of a matrix, and B is electron-phonon coupling constant.

The extreme sensitivity of the ZPL parameters to the local environment of a single-molecule (SM) makes it possible to apply SMSM for the study structure and dynamics of doped solids on the nanometer-scale. In fact, the high ratio of inhomogeneous absorption band width to the very narrow ZPL homogeneous spectral width gives us the opportunity to measure fluorescence excitation spectra and determine coordinates for all sufficiently bright SMs in doped bulk solid samples (crystals, glasses, polymers) at low-temperatures [3-6]. For this purpose, we use specific algorithms and software for rapid SM spectra and images

* Corresponding author: golovanova@isan.troitsk.ru

recognition, data processing and statistical analysis [3].

The technique for sequential-parallel detection of ZPLs and the fluorescence images of myriad probe SMs can be applied for the hyperspectral nanodiagnostics of solids [3, 4, 9, 10]. Namely, various local parameters of dye-doped systems can be.

Analyzing the Γ_0 for SMs in different places of the thin film sample one can map the local fields and corresponding effective n of matrix. For this we proceed the data about 2000 terrylene (Tr) SMs in frozen n-hexadecane film at T=1.5 K. For each Tr SM we found the XY coordinates with different accuracy as it depends on the fluorescence intensity of each SM. We calculate the effective values of the refractive index n in the locality of each SM from its excited state lifetime T_1 dependence on n. This calculation implies the use of the proper theory describing the local-field effects and the $T_1(n)$ dependence [11]. The best fits of the $T_1(n)$ dependence using different models was made and the best model is the virtual cavity model, which describe data more accurately [12]. As a result, the distribution of effective local values of the refractive index in the film was obtained.

In order to estimate the zonal volume that we can attribute to the effective local values of parameters for the h-hexadecane matrix we had to select the molecules for which the coordinate error would not exceed a certain limit. The best-positioned Tr molecules were processed in terms of looking at the distance between two SMs versus the contrast of the effective refractive index each SM indicates in the area around it. All possible combinations of best positioned SM pairs were considered. It was shown that similar, or within the positioning error, values of the effective n are typical for SMs separated by a distance no longer than 100 nm on the XY plane. This may serve as a rough estimation of the volume size to which the concept of the effective n responsible for light emission dynamics from SM may be referred to.

Analyzing the $\Delta\Gamma_{e\text{-}phon}$(T) contribution for each SM we can map the local vibrational dynamics and electron-phonon coupling parameters in the layer of a host matrix.

Thus, in the present paper we show the potential of ZPL-based fluorescence excitation SMSM for the characterization of complex solids on the *nm*-scale.

The work is supported from RFBR (16-32-00841 – study of local field effects; 16-52-00102 – electron-phonon coupling study), Russian Science Foundation (14-12-01415 – development of methods for statistical analysis of SMSM-data). The experimental data, taken for processing were obtained [9, 10] in collaboration with Prof. L. Kador and Prof. J. Köhler at the Chair EP-IV of Bayreuth University (Germany).

References

1. T. Basché, W.E. Moerner, M. Orrit, U.P. Wild, *Single-molecule optical detection, imaging and spectroscopy* (Cambridge, 1997)
2. M. Orrit et al., Faraday Discussions **184**, 275 (2015)
3. A.V. Naumov, Physics Uspekhi **56**, 605 (2013)
4. A.V. Naumov, I.Y. Eremchev, A.A. Gorshelev, Eur. Phys. J. D **68**, 348 (2014)
5. T. Plakhotnik, D. Walser, A. Renn, Phys. Rev. Lett. **77**, 5365 (1996)
6. T. Plakhotnik, E. A. Donley, U.P. Wild, Ann. Rev. Phys. Chem. **48**, 181 (1997)
7. T. Plakhotnik, J. Lumin. **83-84**, 221 (1999)
8. E.A. Donley, T. Plakhotnik, J. Chem. Phys. **114**, 9993 (2001)
9. A.V. Naumov et al., Angew. Chemie. Int. Ed. **48**, 9747 (2009)
10. A.V. Naumov et al., Phys. Chem. Chem. Phys. **13**, 1734 (2011)
11. D.V. Kuznetsov, V.K. Roerich, M.G. Gladush, J. Exp. Theor. Phys. **113**, 647 (2011)
12. T.A. Anikushina et al., Faraday Discussions **184**, 263 (2015)

Experimental study of emission Z-pinch spectra in the axial and radial directions at the Angara-5-1 facility

V.V. Alexandrov[1], G.S. Volkov[1], E.V. Grabovsky[1], A.N. Gritsuk[1,*], K.N. Mitrofanov[1], G.M. Oleinik[1], and A.P. Shevelko[2]

[1]State Research Center of the Russian Federation Troitsk Institute for Innovation and Fusion Research, 108840 Troitsk, Moscow, Russia
[2] P.N. Lebedev Physical Institute RAS, 119991 Moscow, Russia

Abstract. Data on the energy, power and spectra composition of the soft x-ray pulse of powerful Z-pinch plasmas in the axial and radial directions in the photon energy range of 0.02 - 2 keV are presented. The data are obtained from the analysis of experimental results on the implosion of cylindrical arrays with a diameter of 1.2 cm and a height of 1.6 cm of tungsten wires diameter of 6 μm, the linear mass of 220 μg/cm at a current in the range of 2.2 – 3.5 MA at the Angara-5-1 facility.

Currently, to create a substance of high energy density using a spatial concentration of energy flows generated by powerful sources of energy - energy drivers. One of these drivers is a powerful source of soft X-ray radiation (SXR) on the basis of Z-pinch discharges with the use of multi-module accelerators. The spatial distribution of the plasma density during the process of implosion stagnation Z-pinch has a significant impact on the current distribution, the dissipation of magnetic energy, the duration and power of soft X-rays. Implosion of multiwire array starts at the local depletion of the material, so it is not uniform in the axial direction and in the azimuth plane.

The experiments were performed at the Angara-5-1 facility, which consists of eight parallel modules. The amplitude of the pulse current reaches 3.5 MA, maximum voltage at the beginning of the water transmission line is 0.8–0.9MV, voltage pulse width at half maximum of ~100 ns. The power and spatial anisotropy of the pinch SXR emission were measured using two identical sets of four vacuum X-ray detectors (XRDs) on the basis of open vacuum photodiodes with different filters [1]. XUV emission spectra of Z-pinch plasma were recorded using the grazing incidence spectrograph GIS-1 (concave grating with a radius of R = 1 m, 600 grooves/mm, W/Re coating, grazing angle of 4°) in the spectral range of 20–400 Å [2].

In the experiments, the power and energy of Z-pinch radiation were obtained in different spectral ranges using XRD-detectors in the axial and radial directions. Anisotropy obtained for these directions is characteristic for all types of loads. The size of the region emitting soft X-rays was determined by this value of the radiation anisotropy. It was found

* Corresponding author: griar@triniti.ru

that the size of the emitting region is ~4–5 mm in the SXR range for cylindrical multiwire arrays at peak X-ray power, which corresponds to approximately two to threefold radial compression of the plasma. The size of the emitting region at the time of stagnation, approximately coincides in the ranges of 20-70 eV and 120-280 eV. At the same time the image obtained by the time-integrated pinhole camera indicates the presence of a considerable amount of radiating plasma surrounding the pinch – "trailing" mass [3].

The spectra of the emitting plasma were obtained with a spatial resolution. Using these spectra spatial and spectral parameters of the "trailing" mass were determined. These parameters play an important role in the dynamics of implosions and energy power of the whole process because part of the current can flow through it. This could reduce plasma energy near the axis of the array, which is converted into SXR. It is shown that the presence of a trailing" mass reduces the spectral density of radiation and changes the characteristics of spectrum. Sizes of the radiation region obtained from the spectra with a radial resolution coincide with those obtained by XRD-detectors from the radiation anisotropy and pinhole camera images. Comparison of experimental and simulated spectra of radiation of tungsten Z-pinch was carried out.

The procedure to restore the power spectral distribution in different photon energy ranges was solved using data from two sets of XRD with various filters. It has been shown that the results of measurement of the spectra in the radial direction, obtained by independent methods (from the XRD-detectors and the GIS-1 spectrograph), coincide. In the experiment, it was found that the plasma emission of Z-pinch tungsten multiwire cylindrical array in a radial direction is more "harder" than in the axial direction. That is in a radial direction main part of energy is emitted in the range of 70 - 300 eV, while in the axial direction radiating energy is distributed more evenly between the ranges 20 - 70 and 70 - 300 eV.

Thus, the distribution of the SXR Z-Pinch source intensity is relatively isotropic in its height and in the azimuthal angle, and the spectral composition of the radiation in the axial direction corresponds to a "softer" spectrum as compared to the radial.

This work was supported by the Russian Science Foundation under Project № 16-12-10487 and in part by the Russian Foundation for Basic Research under Project 15-02-04411.

References

1. A.V. Branitskiy, G.M. Oleinik, Instrum. Exp. Tech. **22**, 480 (2000)
2. A.P. Shevel'ko, D.E. Bliss, E.D. Kazakov, M.G. Mazarakis, J.S. McGurn, L.V. Knight, K.W. Struve, I.Yu. Tolstikhina, T.J. Weeks, Plasma Phys. Rep. **34**, 944 (2008)
3. A.N. Gritsuk, V.V. Aleksandrov, E.V. Grabovskiy, Y. Laukhin, K.N. Mitrofanov, G.M. Oleinik, G.S. Volkov, I.N. Frolov, A.P. Shevel'ko, IEEE Transactions on Plasma Science **41**, 3184 (2013)

Nanoconstructions on the base of double-stranded DNA molecules and their optical properties

S.G. Skuridin[1], Yu.M. Yevdokimov[1], D.P. Chulkov[2], V.M. Gusev[2], O.N. Kompanets[2,*], and F.V. Vereschagin[2]

[1]Engelgardt Institute of Molecular Biology RAS, 119991 Moscow, Russia
[2]Institute for Spectroscopy RAS, 108840 Troitsk, Moscow, Russia

Abstract. Experimental results have been presented on studying optical properties of nanoconstructions formed of orientationally ordered neighbouring double-stranded DNA molecules in the structure of their liquid-crystalline phases and dispersion particles of these phases including ones cured with intercalators.

The aim of this work was to develop a method to determine the optical properties of orientationally ordered structures in the form of particles (nanostructures) of the cholesteric liquid-crystalline dispersion (CLCD) of double-stranded (ds) DNA in PEG-containing water-salt solutions [1] - this method proved to be the method of circular dichroism (CD).

The results of calculation of the optical activity of the ds DNA CLCD particles, carried out in [2], based on the theory of electromagnetic waves absorption with large molecular systems in the model, when the particle is regarded as a set of cells containing a large amount of absorbing radiation chromophores (nitrogen bases of DNA) and representing a single complex with a specific, changing the spiral direction of the polarizability, allowed to determine the value of the "structural" circular dichroism of DNA CLCD particles and its dependence on various physical parameters (a step of helical twist of "quasinematic" layers of neighboring DNA molecules, the particle size, and other parameters), and most importantly – allowed to come to the conclusion, according to which the formation of a spatially ordered CLCD particles structure must inevitably be accompanied by the appearance of an intense band in the CD spectrum. The amplitude of this band depends on the DNA concentration used to prepare the dispersion (in the range of 5 to 30 μg/ml), the size of the particles themselves, the twist angle of layers, the distance between them, and the negative sign of this intensive band directly says that neighboring layers of right-spiral DNA molecules have a left spatial twist.

Applied in [2] approach has also been used to study the features of the CD spectra of DNA CLCD treated with some colored biologically active compounds (BAC), which can intercalated between pairs of DNA nitrogen bases and turn out to be included in the three-dimensional structure of DNA CLCD particles.

* Corresponding author: onkomp@isan.troitsk.ru

Design data were compared with the experimental results obtained by studying the influence of various factors on the DNA CLCD particles including BAC interacting with ds DNA. The optical properties in the UV region were examined for the samples of DNA CLCD and the same DNA dispersion treated with daunomycin (DAU) – "coloured" anthracycline antibiotic used in the treatment of many malignancies. Registration of the CD optical signal was performed with the CDS-2MUV dichrometer (the ISAN development) adapted for use with DNA nanostructures.

The experimental CD spectra showed that the formation of the DNA dispersion is accompanied by appearance of abnormal negative band in the CD spectrum (with a minimum at $\lambda \sim 270$ nm), located in the region of the absorption of nitrogenous bases, with the concurrence of its form to theoretically calculated one that leads to the unambiguous conclusion about the liquid-crystalline cholesteric packaging of DNA molecules in the dispersion particles - a particular property of molecules having geometrical and optical anisotropy.

The CD spectrum of DNA CLCD "coloured" by DAU comprises two abnormal bands having negative signs: one - in the absorption band of DNA nitrogen bases (at $\lambda \sim 270$ nm), and the other - in the absorption band of DAU chromophores (at $\lambda \sim 500$ nm). Coincidence of abnormal bands signs arranged in different regions of the spectrum suggests that the DAU molecules intercalated between DNA base pairs with 90 degree inclination angle with respect to the long axis of the DNA molecule. If the amplitude of the abnormal band in the absorption region of the DNA nitrogen bases remains practically unchanged, the amplitude of the abnormal band in the absorption region of DAU molecules increases with the degree of their DNA binding. In fact, this relationship is a calibration curve (Fig. 1), which can be used to determine the DAU concentration in the DAU-containing sample solution, and the DNA CLCD itself is a biosensing unit that sensitively reacts to the BAC content in the solution.

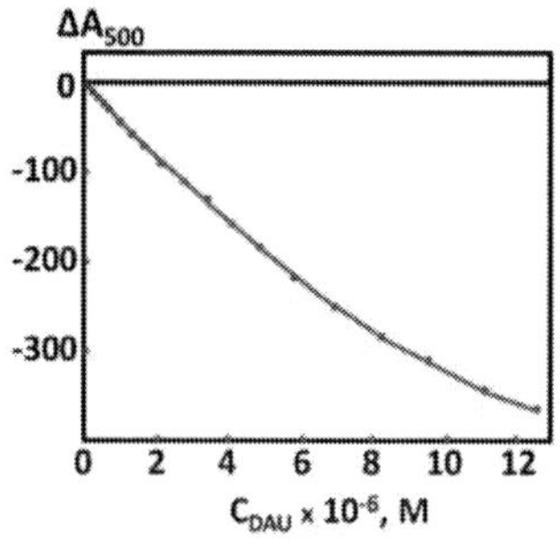

Fig. 1. The dependence of the amplitude of the 500 nm abnormal band in the CD spectrum of the CLCD particles of the complex (DNA-DAU) on the concentration of added DAU. $C_{DNA} = 10$ µg/ml; $C_{PEG} = 170$ µg/ml; $\Delta A_{500} = (A_L - A_R) \times 10^{-6}$ o.u.; L = 1 cm.

The support from the Program of RAS Presidium "Fundamental and applied problems of photonics and optical physics of new materials" is acknowledged.

References

1. Yu.M. Yevdokimov, V.I. Salyanov, S.G. Skuridin, *Nanostructures and nanoconstructions based on DNA* (CRC Press, Boca Raton, 2012)
2. Yu.M. Evdokimov, S.V. Semenov, S.G. Skuridin, V.I. Salyanov, P.N. Solev, V.T. Valuev - Elliston, F.V. Vereshchagin, D.P. Chulkov, O.N. Kompanets, Russian: Sensor Systems **30**, 249 (2016)

DOI: 10.1051/epjconf/201713203011

Calculation of IR-spectra of structural fragments of lignins

O.Yu. Derkacheva*, and M.M. Ishankhodzhaeva

St. Petersburg State University of industrial technologies and design, 198095 St. Petersburg, Russia

Abstract. To study structure of softwood lignins the experimental and theoretical IR-spectra in middle IR-diapason were analyzed. To interpret these data the quantum chemical calculations of IR-spectra of general dimmer fragments of softwood lignins by method of density functional theory (DFT/B3LYP) with 6-31G(d,p) as basis set were carried out. These calculations showed that frequencies of normal vibrations of fragment with β-alkyl-aryl linkage are close to the experimental values of the IR absorption bands of lignin, and infrared spectrum of this structure is similar to the experimental spectrum of lignin. The calculations with accounting for the solvent showed a strong increase in the intensity of the majority of the bands and the solvent effect on the frequencies of vibrations.

Lignins, being part of the plant raw, are valuable source of chemical compounds [1]. It is why heightened interest to decoding native lignin structure is saved. Lignin is not individual substance. Lignin is mixture of irregular aromatic polymers of close structure. Therefore the structural formula of lignin does not exist.

At the same time it is known that lignins consist of phenylpropane units of guaiacyl, syringyl and p-coumaryl types [2]. Currently it is revealed more than 20 types of linkages between phenylpropane units. It is established that β-alkyl-aryl linkage is the most common one in native lignins.

One of useful methods of lignin analysis is FTIR-spectroscopy. By quantitative analysis of IR-spectra and construction of calibration dependencies similar "structure-properties" the tasks of finding the most informative absorbance bands to characterize chemical structure of native lignins are quite successfully solved [3-7]. In order to increase the reliability of the interpretation of IR spectra there is a necessity of the comparative analysis of the experimental and theoretical data. To achieve this aim in the work the quantum-chemical calculations of IR spectra of dimmer models of softwood lignin were carried out by method of density functional theory with the B3LYP functional in the basis of 6-31 + G (d, p) using a software package Gaussian08. IR-spectra were calculated for two main structural elements of guaiacyl lignin.

Figure shows the lignin experimental spectrum and the simulated IR spectra of the investigated dimmers in the form of the absorption curves in the frequency range 700 – 1900 cm^{-1}. By comparison with experiment the frequencies of normal vibrations of the

* Corresponding author: derkachevaou@rambler.ru

dimmers were multiplied by a scaling factor 0.975 for the frequency range 400 – 1900 cm^{-1} and 0.950 for the region 2700 – 4000 cm^{-1} [8-9].

The calculation results showed that the frequencies of normal vibrations of the dimmer with β-O-4 linkage between aromatic rings are close to the experimental values of the infrared absorption bands of lignin, and the calculated infrared spectrum (curve 2 a) is similar to the experimental spectrum of lignin (curve 1).

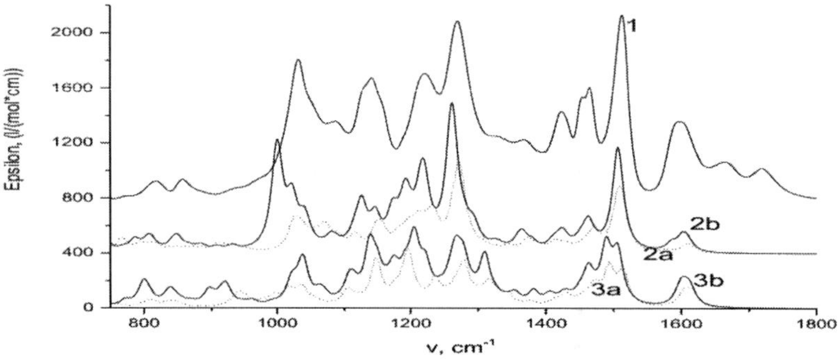

Fig. 1. IR-spectrum of softwood lignin (1) and molar absorption coefficient of dimmers with β-O-4 (2) and with α-O-4 и β-5 linkages (3) in vacuum (a) and in water (b)

IR-spectrum of model with α-O-4 and β-5 linkages between aromatic rings (curve 3a) is markedly different from spectrum of the previous model. We can note, for example, the difference between the frequencies of in-plane skeletal vibration of two aromatic rings of the dimmer near 1500 cm^{-1}, which is the cause of an appearance of doublet with frequencies at 1498 and 1517 cm^{-1}. Quantum chemical calculations showed a marked increase in the intensities of the bands at 1150 and 1200 cm^{-1} and the appearance of strong bands at 1320 and 940 cm^{-1}. According to calculations, these bands are associated with in-plane vibrations of aromatic ring with four substitutes.

In the work an influence of environment on the spectra of both dimmers was considered. To see a solvent effect (water) the calculations were carried out by method of PCM (model of polarization continuum Tomasi). From Fig. 1 we can see that the IR-spectra calculated by quantum-chemical method of PCM (curves 2b and 3b) demonstrate an increasing intensity of dimmer absorbance bands as result of environment influence. One can also note shift of absorbance bands of aromatic rings on 7 – 10 cm^{-1} to lower frequencies.

Thus comparing theoretical IR-spectra of the studied structural fragments with experimental spectra of softwood lignins allows to clarify an empirical assignment of absorbance bands. Later it will allow to evaluate structure of lignins of different origin.

References

1. I. P. Deimeko, Khimija Rastitel'nogo Syr'ja **16**, 5 (2012) (in Russian)
2. S.Y. Lin, C.W. Dence, *Methods in lignin chemistry* (Springer-Verlag, Berlin, 1992)
3. K.V. Sarkanen, H.M. Chang, G.G. Allan, Tappi J. **50**, 587 (1967)
4. A.P. Karmanov, O.Yu. Derkacheva, Russ. J. Bioorganic Chem. **39**, 677 (2013)
5. O.Yu. Derkacheva, Journal of Applied Spectroscopy **80**, 670 (2013)
6. N.N. Chipanina et al., Chemistry of Natural Compounds **31**, 392 (1995)
7. O. Faix, Holzforschung **40**, 273 (1986)
8. A.P. Scott, L. Radom, J. Phys. Chem. **100**, 16502 (1996)
9. S. Barsberg, Theor. Chem. Acc. **134**, 33 (2015)

Structural and optical properties of $Nd_xGd_{1-x}Cr_3(BO_3)_4$ solid solutions

E.A. Dobretsova[1,2,*], K.N. Boldyrev[1], E.Yu. Borovikova[3], and V.A. Chernyshev[4]

[1]Institute of Spectroscopy RAS, 108480 Troitsk, Moscow, Russia
[2]P.N. Lebedev Physical Institute RAS, 119991 Moscow, Russia
[3]M.V. Lomonosov Moscow State University, 119991 Moscow, Russia
[4]Ural Federal University, 620002 Ekaterinburg, Russia

Abstract. The spectroscopic investigation of the solid solutions $GdCr_3(BO_3)_4$ (sp.gr. *R*32) – $NdCr_3(BO_3)_4$ (sp.gr. *C*2/*c*) is presented. The single crystals $Nd_xGd_{1-x}Cr_3(BO_3)_4$, $0 \leq x \leq 1$ with the step of 0.1, were grown by a flux method. Crystal structures were identified by terahertz and optical spectroscopy methods. *Ab initio* calculations in the frame of density functional theory allowed us to separate modes belonging to the *R*32 and *C*2/*c* phases and to estimate the ratio of these phases in $Nd_xGd_{1-x}Cr_3(BO_3)_4$ crystals. Content of the monoclinic phase (C2/c) increases with growth of the Nd concentration.

The borate family with a general formula $RM_3(BO_3)_4$ (*R* stands for Y or rare-earth (RE) elements, *M* = Al, Fe, Cr, Ga, Sc, Mn) has wide variety of physical properties depending on chemical composition and growth conditions. The borate crystals are well known as functional materials, and they are thermally and chemically stable. They are perspective for applications due to their optical, magnetic and magnetoelectric properties. Magnetic properties of RE borates with the magnetic ions Fe^{3+} and Cr^{3+} are of fundamental interest due to a coexistence and interaction of two magnetic subsystems (3*d* and 4*f* ions). Moreover, interaction in the iron (chromium) subsystem has quasi-one-dimensional character. Though there is a lot of research on the iron borates, the RE chromium borates are still understudied [1-3]. The analysis of thermal, magnetic, and optical properties of $NdCr_3(BO_3)_4$ has shown that the compound orders antiferromagnetically into an easy-plane magnetic structure at the temperature T_N=8 K [1]. The RE chromium borates have polytypic nature and, depending on growth conditions, crystallize in two modifications, namely, a non-centrosymmetric rhombohedral one with the space group *R*32 and a centrosymmetric monoclinic one with the space group *C*2/*c* [4, 5]. Coexistence of differently ordered fragments in a single crystal is a characteristic feature of these compounds.

In this paper, we show results of solid solutions $GdCr_3(BO_3)_4$ (sp.gr. *R*32) – $NdCr_3(BO_3)_4$ (sp.gr. *C*2/*c*) investigation by terahertz and optical spectroscopy methods. The purpose of our investigation was to trace the behaviour of the borate crystal structure at the substitution of Gd by Nd. On the ground of the results on the iron borates [6], we suppose

* Corresponding author: elena-dobrecova@yandex.ru

that the compound with maximum content of Nd and primary rhombohedral non-centrosymmetric crystal structure will demonstrate a maximum magnetoelectric response.

We have measured absorbance spectra of $Nd_xGd_{1-x}Cr_3(BO_3)_4$ borates in the range of the $^4I_{9/2} \rightarrow {}^4F_{3/2}$ electronic transition in the Nd^{3+} ion. For the samples with x<0.6, at temperature T=16 K> T_N, we observe two spectral lines (corresponding to the transitions from the ground Kramers doublet to the two CF levels of $^4F_{3/2}$). For x>0.6, the lowest-frequency spectral line splits into two components. Because the temperature of magnetic ordering of the chromium borates is lower than 16 K [1], this splitting is not related to a splitting of Kramers doublets in an internal magnetic field. Thus, in the case of Nd^{3+} ion concentration more than 60% in the solid solutions, the frequencies of this line for the rhombohedral and monoclinic phases are different. We can estimate a structural phases' ratio in the crystals. The more intensive line represents the monoclinic phase and the less intensive one – the rhombohedral phase.

The terahertz range of IR spectra displays vibrations of the RE ions. The main change when changing x is in intensity redistribution between bands with frequencies ~ 70 cm^{-1} and 77 cm^{-1} related to the monoclinic and rhombohedral phases, respectively. In the spectrum of $GdCr_3(BO_3)_4$, the band 77 cm^{-1} has a shoulder 70 cm^{-1} which is due to a small admixture of the monoclinic phase. Decomposition of the spectrum into two Gaussians shows that the intensity of the stronger band is above 80%. According to the *ab inito* calculations, the oscillator strength of the vibration in the monoclinic phase is 1.5 times larger than that of the rhombohedral one. Considering this, we estimate the rhombohedral phase content as about 85%. We observe a large content of the monoclinic phase (above 40% taking into account the oscillator strengths) already in the case of 20% Nd concentration in the crystals. If the Nd concentration is more than 70%, the low-frequency bands of the rhombohedral and the monoclinic phases coincide. For this reason, it is impossible to distinguish the structural phases in terahertz range for these compositions.

Thus, optical and terahertz spectroscopy methods complement each other in determination of the structural phase ratio in the $Nd_xGd_{1-x}Cr_3(BO_3)_4$ solid solutions. Based on the obtained data, we can claim that the $Nd_xGd_{1-x}Cr_3(BO_3)_4$ borates with x≤0.2 have essentially rhombohedral non-centrosymmetric structure (sp.gr. $R32$). This points to a possible existence of a considerable magnetoelectric effect in these compounds, the same as in non-centrosymmetric RE iron, aluminum and gallium borates. The next stage of the work is to prepare large crystals of the solid solutions with x≤0.2 for checking the magnetoelectric response.

This work was supported by the Russian Science Foundation (Grant No 14-12-01033). E.A.D. acknowledges a Scholarship of the President of the Russian Federation (SP-754.2015.1).

References

1. E.A. Popova et al., Phys. Rev. B **75**, 054446 (2007)
2. K.N. Boldyrev, E.P. Chukalina, N.I. Leonyuk, Phys. Sol. St. **50**, 1681 (2008)
3. L. Gondek et al., J. Sol. St. Chem. **210**, 30 (2014)
4. V.S. Kurazhkovskaya et al., J. Struct. Chem. **52**, 699 (2011)
5. E.Yu. Borovikova et al., Vibr. Spectr. **68**, 82 (2013)
6. A.M. Kadomtseva et al., Low Temp. Phys., **36**, 511 (2010)

Holmium iron borate: high-resolution spectroscopy and crystal-field parameters

D.A. Erofeev[1,2,*], E.P. Chukalina[1], M.N. Popova[1], B.Z. Malkin[3], L.N. Bezmaternykh[4], and I.A. Gudim[4]

[1]Institute for Spectroscopy RAS, 108840 Troitsk, Moscow, Russia
[2]Moscow Institute of Physics and Technology, 141700 Dolgoprudny, Russia
[3]Kazan Federal University, 420008 Kazan, Russia
[4]Kirensky Institute of Physics Siberian Branch RAS, 660036 Krasnoyarsk, Russia

Abstract. High-resolution transmission spectra of $HoFe_3(BO_3)_4$ single crystals were measured in broad spectral (5000-23000 cm^{-1}) and temperature (1.7-300 K) ranges. Crystal-field energies of the Ho^{3+} ions were determined for a paramagnetic and easy-axis antiferromagnetic phases of the compound. On the basis of these data and of preliminary crystal-field calculations in the frame of the exchange-charge model, crystal-field parameters were found. A parameter of the isotropic Ho-Fe exchange interaction was estimated.

Holmium iron borate belongs to the family of new multiferroic materials – rare-earth (RE) borates with a trigonal structure of the mineral huntite. It crystalizes in the *R*32 symmetry space group (SG). The compound exhibits a substantial magnetoelectric effect at temperatures below T_N = 39 K where an antiferromagnetic ordering into an easy-plane magnetic structure takes place. To interpret magnetoelectric properties of RE compounds one needs the information on RE ion crystal-field (CF) levels and wave functions [1]. Detailed research on CF levels of Ho^{3+} in $HoFe_3(BO_3)_4$ has not been carried out before. The task is complicated by the fact, that the crystal has a low-symmetry structure with the SG $P3_121$ at room temperature, while the structural phase transition $R32 \rightarrow P3_121$ occurs at $T_s\sim$ 360 K [2]. The point symmetry group of the holmium site in the $P3_121$ structure of $HoFe_3(BO_3)_4$ is C_2. Below T_{SR} = 5 K, the magnetic structure of holmium iron borate changes to an easy-axis one.

High-quality $HoFe_3(BO_3)_4$ single crystals were grown from fluxes based on $Bi_2Mo_3O_{12}$. The transmission spectra of oriented single crystals were registered using a high-resolution Fourier-transform spectrometer Bruker IFS 125HR in the spectral range 5000–23000 cm^{-1} with a resolution up to 0.2 cm^{-1}, which is sufficient to detect all spectral details. The crystal sample was placed into either a closed-cycle cryostat Cryomech ST403 (3.5 – 300 K) or a helium-filled optical cryostat with helium vapor pumping (1.7 – 4.2 K). The analysis of temperature dependences of polarized light absorption allowed us to identify the crystal-field levels of the Ho^{3+} ion in the paramagnetic phase of $HoFe_3(BO_3)_4$. In particular, the energies of CF levels of the ground multiplet $^5I_8$ are 0, 7.5, 14, 18, 54, 66, 95, 137, 154,

* Corresponding author kinson@mail.ru

175, 207, 246, 272, 311, 322, 340, and 354 cm^{-1}. The energies of upper levels from this dataset were specified using cooperative satellites in the high-frequency part of the absorption spectra. These satellites correspond to a simultaneous excitation of the two interacting holmium ions by one photon [3]. The knowledge of CF levels of the ground multiplet is important for interpretation of magnetic and thermodynamic properties of the compound. We have also determined the CF energies of Ho^{3+} ions in the easy-axis phase of the crystal ($T<T_{SR}$). As for the easy-plane phase, more work is necessary to identify all the observed spectral lines in a very complicated spectrum in the temperature range $T_{SR} <T< T_N$.

The energy spectrum of the Ho^{3+} ion in the crystal field of the C_2 symmetry in $HoFe_3(BO_3)_4$ is defined by 15 parameters B_q^p in the CF Hamiltonian, which is represented by a linear combination of spherical tensor operators $C_q^{(p)}$ in the Cartesian coordinate system with $x||a$ and $z||c$ axes. The parameters calculated with the exchange-charge model [4] were used as initial parameters. Then, the energies of transitions between the levels of the Ho^{3+} ions, obtained from a numerical diagonalization of the complete Hamiltonian operating in the space of 1001 states of the $4f^{10}$ electronic configuration, were compared with the measured optical spectra of $HoFe_3(BO_3)_4$ in the paramagnetic phase, and the initial CF parameters were varied to fit the experimental data. The following set of the CF parameters (in cm^{-1}) was obtained: $B_0^2 = 376$, $B_0^4 = -1112$, $B_{-3}^4 = B_3^4 = 526i$, $B_0^6 = 336$, $B_{-3}^6 = B_3^6 = 65i$, $B_6^6 = B_{-6}^6 = 232$ (they define the trigonal component of the crystal field); $B_{-1}^2 = B_1^2 = 71i$, $B_2^2 = B_{-2}^2 = 65$, $B_{-1}^4 = B_1^4 = -52i$, $B_2^4 = B_{-2}^4 = 71$, $B_4^4 = B_{-4}^4 = -16$, $B_{-1}^6 = B_1^6 = -21i$, $B_2^6 = B_{-2}^6 = -9.6$, $B_4^6 = B_{-4}^6 = -27$, $B_{-5}^6 = B_5^6 = -70i$ (these parameters are responsible for the low-symmetry C_2 component).

All the ions of holmium in the paramagnetic and easy-axis antiferromagnetic phases of $HoFe_3(BO_3)_4$ are optically equivalent in the absence of an external magnetic field. The frequency shifts of the electronic transitions (shifts of energy levels in the exchange field) observed between the temperatures T=50 K > T_N and T = 1.5 K < T_{SR} were interpreted considering the isotropic exchange interaction between the Ho^{3+} ions and Fe^{3+} ions in the *S* state, $H_{\text{RE-Fe}} = -J\boldsymbol{S}_{Ho} \cdot \boldsymbol{S}_{\text{Fe}}$ ($\boldsymbol{S}_{Ho}$ и $\boldsymbol{S}_{Fe}$ are spin moment operators for the holmium and iron ions). The value of the isotropic exchange interaction parameter $J = 0.2$ cm^{-1} was determined from a comparison of calculated spectral line shifts with the experimental data.

This work was supported by the Russian Science Foundation (Grant № 14-12-01033).

References

1. A.I. Popov, D.I. Plokhov, A.K. Zvezdin, Phys. Rev. B **87**, 024413 (2013)
2. D.A. Erofeev., E.P. Chukalina, L.N. Bezmaternykh, I.A. Gudim, M.N. Popova, Opt. Spectrosc. **120**, 558 (2016)
3. A.P. Abramov, N.I. Agladze, I.Ya. Gerlovin, M.N. Popova, Opt. Spectrosc. **64**, 1042 (1988)
4. B.Z. Malkin, *Ion-phonon interactions. Spectroscopic Properties of Rare Earths in Optical Materials* (Springer, Berlin, 2005)

Positron annihilation spectroscopy on a beam of positrons the LEPTA facility

E.V. Ahmanova[1], M.K. Eseev[2,*], A.G. Kobets[1,3], I.N. Meshkov[1,4], O.S. Orlov[1], A.A. Sidorin[1], K. Siemek[1,5], and P. Horodek[1,5]

[1]Joint Institute for Nuclear Research, 141980 Dubna, Russia
[2]M.V. Lomonosov Northern (Arctic) Federal University, 163002 Arkhangelsk, Russia
[3]Institute of Electrophysics and Radiation Technologies NAS of Ukraine, 61002 Kharkov, Ukraine
[4]Saint Petersburg State University, 199034 St. Petersburg, Russia
[5]Institute of Nuclear Physics Polish Academy of Sciences, 31-342 Krakow, Poland

Abstract. The results and possibilities of the samples surfaces research by the Doppler method of positron annihilation spectroscopy (PAS) for a monochromatic beam of positrons at the LEPTA facility are presented in this paper. Method with high-resolution sensitivity to defects like vacancies and dislocations allows scanning of the surface and near-surface sample layers to a depth of several micrometers by the method of Doppler broadening of annihilation lines. The opportunities for the development of a PAS method based on the measurement of the positron lifetime in the sample irradiated by ordered flow of positrons from the injector of accelerator complex LEPTA at JINR are discussed.

LEPTA facility [1] is a complex consisting of the cryogenic positron source based on the emitter - a radioactive 22Na with an activity of 30 mCi, the electromagnetic Penning-Malmberg-Surko trap (PMS) and storage ring. After slowing down in a frozen solid neon layer a monochromatic flux of positrons with intensity up to 10^6 part./sec is formed. The width of the positron spectrum is about 2 eV. This flux can be used for fundamental and applied research [2]. It is possible to carry out layer-by-layer scanning of the defects in subsurface layers of the samples at depths from zero to several micrometers by varying the positron energy in the range of 0.05÷35 Kev. For this purpose, one of three versions of the PAS can be used [3]: observation of correlations in the angular distribution of annihilating photons – gamma quanta (ADAP), measurement of the lifetime of positrons in matter with registration the time distribution of the annihilation photons (TDAP) and the change in energy of the annihilation gamma-quanta - registration of Doppler broadening of annihilation lines (DBAL).

The most perspective methods that allow to distinguish the defects of vacancy and dislocation types in samples with a sensitivity of 1 defect per 10^7 atoms in the crystal lattice and at depth resolution of the sample in tenths of nanometer, are the last two versions - TDAP and DBAL. Currently, at the LEPTA complex DBAL PAS method is used with the

* Corresponding author: m.eseev@narfu.ru

positron beam to successfully carry out defectoscopy of different materials: the steel samples and others. [4, 5, 6].

Currently the dedicated channel with camera for positron annihilation spectroscopy is mounted. It will improve the technology of the samples study. Recently a helium cryocooler was installed that simplifies significantly operation of the formation of a layer of solid neon on the surface of ^{22}Na tablet to monochromatize the beam of positrons from the source. The method TDAP PAS for layer-by-layer scanning of the defects with a monochromatic beam of positrons is under development as well. Two implementations of TDAP are considered: controlled positron beam and operating the Penning-Malmberg-Surko trap at LEPTA, allowing to create bunches of positrons of high intensity [7, 8]. In the latter case, for realization the TDAP method can be used a bunching system for positron accumulate in the trap will be used (similar to the AEgIS experiments at CERN [9]). The bunch will be subsequently accelerated to achieve the required time resolution.

References

1. E.V. Akhmanova, M.K. Eseev, A.G. Kobets, I.N.Meshkov, A.Y. Rudakov, A.A. Sidorin, S.L. Yakovenko, Phys. Part. Nucl. Lett. **9**, 373 (2012)
2. A.A. Sidorin, I. Meshkov, E. Ahmanova, M. Eseev, A. Kobets, V. Lokhmatov, V. Pavlov, S. Yakovenko, Materials Science Forum **733**, 322 (2013)
3. M.J. Puska, R.M. Nieminen, Rev. Mod. Phys. **66**, 841 (1994)
4. P. Horodek, K. Siemek, A.G. Kobets, M. Kulika, I.N. Meshkov, Appl. Surf. Sciense **333**, 96 (2015)
5. P. Horodek, M.K. Eseev, A.G. Kobets, Nukleonika **60**, 721 (2015)
6. E. Ahmanova, P. Horodek, I. Meshkov, O. Orlov, A.A. Sidorin, A. Kobets, M. Eseev, *Proceedings of the International Workshop on Beam Cooling and Related Topics* (Newport News, USA, 2015)
7. M.K. Eseev, A.G. Kobets, I.N. Meshkov, A.A. Sidorin, O.S. Orlov, JETP Letters **102**, 261 (2015)
8. M.K. Eseev, I.N. Meshkov, Physics Uspekhi **59**, 304 (2016)
9. S. Aghion et. al., Nucl. Instr. Meth. Phys. Res. Sect. B **362**, 86 (2015)

EPJ Web of Conferences **132**, 03015 (2017) DOI: 10.1051/epjconf/201713203015
SPECTROSCOPY.SU 2016

Investigation of multilayer painting by means of molecular spectroscopy

I.F. Kadikova[1,*], and I.A. Grigorieva[2]

[1]The State Research Institute for Restoration, 107014 Moscow Russia
[2]The State Hermitage Museum, 190000 Saint Petersburg Russia

Abstract. The study of paintings' layer structure under microscope was implemented in the early XX century; since that time method of cross-section preparation as well as composition of embedded materials changed and capabilities of optical microscopy broadened. Nevertheless, until recently cross-sections were prepared only for analysis of paintings' layer-specific structure, morphological and technological features, tracks of restoration treatments. Examination of chemical composition required additional sampling as well as division of these samples layer by layer in order to determine pigments and binding media; these procedures didn't always lead to proper results. Micro-Raman and micro-FTIR spectroscopy allow carrying out investigation of painting cross-sections without any additional preliminary sampling; this is shown on example of investigation of a number of paintings.

For the first time painting stratigraphy issue attracted investigators' attention in the early XX century, when in 1914 paintings' structure was studied under microscope on a picture and on painting samples put on edge. Later, while technology advanced the method improved, different variants of embedded materials for cross-sections preparation were suggested and capabilities of optical spectroscopy developed [1]. But the procedure of cross-section preparation remains the same and consists in the following: painting sample of 1-1.5 mm size which contains as many layers as possible (ground, paint layers and varnish) is placed in an embedded material at 90° angle.

Until recently microscopic investigation of painting cross-sections allowed determining only the technological features of a picture: paint layer structure, quantity, colour and thickness of layers, presence of imprimatura and underpainting, tracks of restoration interventions. Determination of pigment and binding media composition in painting is a difficult issue because components of each layer should be investigated using different analytical methods including optical microscopy, X-ray fluorescence analysis, electronic microscopy, molecular spectroscopy, gas chromatography etc. [2, 3]. For example, thanks to investigation of cross-sections from different parts of the picture of unknown artist (F. Budkin, "Portrait of M. Lermontov") three primers were revealed, which is a rare occurrence. Also, study of icon "Hagiography of St. Parthenius, Bishop of Lampsacus" (Uspenskiy

* Corresponding author: kadikovairina@mail.ru

Cathedral, Smolensk) established that latter restoration wispy paint layer drawn above the original painting in fact has an underlying calcium carbonate ground.

Most conventional FTIR spectroscopy methods are invasive and require relatively large amount of sample. For a good absorbance spectrum an additional sampling is needed: it consists in mixing of a sample and potassium bromide with further grinding and pressing into pellets. ATR spectroscopy also have serious drawbacks: samples can not be used in further investigations due to changing of their layer structure after contact with ATR crystal and the method provide the information regarding only the upper layer. Unlike methods described above, micro-Raman and micro-FTIR spectroscopy allow examining paint layer composition directly from a cross-section. That spares a sample and makes it possible to use it in further investigation.

One of the advantages of micro-FTIR technique is capability of collecting spectra in different modes, among which the most suitable for cross-sections investigation is ATR mode. It requires small amount of a material, minimal sample preparation process and is able to achieve higher spatial resolution (about 10 μm) due to higher refractive index of the ATR crystal (Ge). It is possible to obtain high-quality IR spectra of most of paint layers with thickness no less than 10 μm. In case of investigation of thinner layers precision of analysis can be maintained by changing the configuration and aperture size. Micro-Raman spectroscopy which has higher spatial resolution (about 1 μm) allows obtaining spectra from separate grains of pigments and minerals, determining structural features of different pigments, for example, defining crystal system of titan white (TiO_2).

Complementary nature of Raman and IR spectroscopy can be illustrated by analysis of a cross-section from Titian painting. Verdigris was found in the paint layer by means of micro-ATR-FTIR spectroscopy, and lead-tin yellow type I was found with the help of micro-Raman spectroscopy.

Raman and ATR-FTIR micro-spectroscopy are valuable analytical techniques for analysis of paint cross-sections. Both methods require minimal sample preparation and have high spatial resolution. Besides, the technique allows performing chemical mapping – generation of visual images depicting the distribution of components of paint layers, variations of molecular structure including the result of paint layers degradation.

References

1. M.J. Wachowiak, JAIC **43**, 205 (2004)
2. J.S. Tsang, JAIC **30**, 163 (1991)
3. B. Singer, T.E. Aslaksby, B. Topalova-Casadiego, Stud. Conserv. **55**, 274 (2010)

Frequency stabilization of diode laser on the wavelength of $5P_{3/2} \rightarrow 5D_{5/2}$ of rubidium transition

A.Yu. Kalatskiy[1,2], A.E. Afanasiev[1,*], P.N. Melentiev[1], and V.I. Balykin[1]

[1]Institute for Spectroscopy RAS, 108840 Moscow, Troitsk, Russia
[2]Moscow Institute of Physics and Technology, 141700 Dolgoprudny, Russia

Abstract. The method of frequency stabilization of diode laser on the wavelength of 5P3/2→5D5/2 of rubidium (776 nm) has been realized. The two-photon absorption spectroscopy of the rubidium vapors is underlie of the suggested scheme. The basic parameters such as the intensity of laser radiation and concentration of the vapors have been investigated. The frequency stability of radiation was about 1 MHz

There are several approaches to the stabilization of the laser radiation at the wavelength of 776 nm, which corresponds to the second transition 5P to 5D of rubidium atom (the first is 5S to 5P). One of the simplest approaches is based on the use of a high-finesse and a highly stable Fabry–Perot etalon in combination with a wavelength meter (lambda meter). In this scheme, the laser radiation is stabilized with respect to the transmission peak of the Fabry–Perot etalon [1], and the stabilization frequency can be chosen arbitrarily. The lambda meter is used to determine the wavelength of the laser radiation. The main problem of this scheme is the frequency drift of the Fabry–Perot etalon, which restricts the stabilization time of the laser radiation.

Another approach is based on the use of resonant absorption lines of the atom. Because the frequency of any atomic transition is stable in time, this approach does not have fundamental physical restrictions on the long-term stability of the laser. In this approach to the frequency stabilization of the laser radiation, the characteristic stabilization time is determined only by the drift of electronic components of the laser source and stabilization system.

The frequency stabilization of a laser on an atomic transition 5P to 5D can be implemented by different methods. Here, three main approaches should be singled out, which are based on the use of (1) the electromagnetically induced transparency (EIT) [2] of the first atomic transition (5S to 5P); (2) the fluorescence due to the decay of the second excited state (5D); (3) the absorption of radiation on the second transition (5P to 5D). All the three methods utilize the stepwise excitation scheme 5S to 5P to 5D. In all these cases, on the first transition (5S to 5P) the laser is stabilized by using standard methods based on the saturated absorption transition with implementation of lock-in technique [3] or stabilisation on the slope of the nonlinear absorption resonances [4].

* Corresponding author: afanasiev@isan.troitsk.ru

The experimental scheme of frequency stabilization of a diode laser on the 5P to 5D transition of the Rb atom using the absorption scheme is presented on figure 1. The scheme is based on the two-step spectroscopy of Rb atoms in a field of two counterpropagating laser beams. For its realization, we used two laser systems, a cell with vapors of Rb atoms, photodiodes, and accessory optical elements.

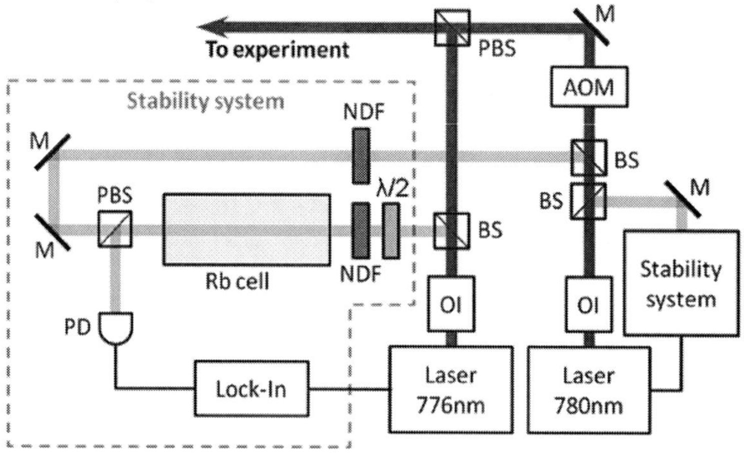

Fig. 1. Schematic of the frequency stabilization system of a diode laser operating on the transition 5P to 5D of Rb atom: (1) filter, (2) λ/2 plate, (3) beam-splitting cube, (4) mirror, (5) photodiode, (6) polarizing cube, (7) cell with Rb vapors, (OI) optical insulator, and acousto-optic modulator (AOM).

The measurements of the laser frequency stability at the wavelength of 776 nm has been realized with atomic beam of Rb in vacuume chamber. Without stabilization, the laser goes out of resonance with the atomic beam in a time of about 5 min (shift about 100 MHz). The stabilized laser retains its frequency for a few hours. The obtained stability of the laser is about 1 MHz, which is comparable with the stabilization accuracy of the laser on the first transition (780 nm).

This work was supported by the Program of the Presidium of the Russian Academy of Sciences.

References

1. Y.C. Chung, T.M. Shay, Optical Engineering **27**, 275424 (1988)
2. K.J. Boller, A. Imamoglu, S.E. Harris, Phys. Rev. Lett. **66**, 2593 (1991)
3. M. Weel, A. Kumarakrishnan, Canadian Journal of Physics **80**, 1449 (2002)
4. P. Borisov, P. Melentiev, S. Rudnev, V. Balykin, Las. Phys. **15**, 1 (2005)

Spectrometric determination of clinically relevant fatty acids in the blood serum

A.V. Kalinin[1,*], V.N. Krasheninnikov[1], A.P. Sviridov[2], and V.N. Titov[3]

[1]Institute for Spectroscopy RAS, 108840 Troitsk, Moscow, Russia
[2]Institute on Photon Technologies RAS, 108840 Troitsk, Moscow, Russia
[3]Russian Cardiology Research and Production Complex, 119121 Moscow, Russia

Abstract. The content of fatty acid (FA) triglycerides in food and biological media is predicting traditionally using gas and liquid chromatographic methods. Named techniques aren't available for clinical labs due to their complexity. So, our objective was to develop the method and apparatus for rapid assay of a few clinically important FA as the saturated palmitic, mono unsaturated oleic and others in serum using near infrared spectrometer. As a result, the applicability of the FT spectrometer in the wavelength range of 0.9 -1.8 μ to analyze these FA in serum without sample preparation was confirmed. Besides, measurement specifications were determined and a correlations of the absorption spectra and contents of total triglycerides and cholesterol, palmitic, oleic, linoleic and arachidonic FA in serum were established

It is well known [1,2] that cardiovascular diseases and cancer - the main causes of human death, are statistically associated with an increased content of saturated fatty acid triglycerides (FA) compared to unsaturated FA in the blood. The most significant FA should include saturated palmitic (C16:0) and unsaturated oleic (C18:1) and linoleic (C18:2) FA. When treating cardiovascular diseases there is an urgent need for rapid methods of determining the content of these FA in blood. Traditional methods of FA analyses are the gas and liquid chromatography, they are of little use for express applications in clinical labs because of inefficacy, the complexity of sample preparation and high cost of analysis. The usage of near infrared spectroscopy, combined with the method of multi-component calibration - regression on projections to latent structures (PLS), seems promising for mass express-control of the above-mentioned FA in serum of patients [3].

Spectrometric analysis of FA triglycerides includes laborious calibration of the spectrometer on a huge number of standard samples with varying concentrations of several components. For the present study three sets of 27, 30 and 42 serum samples taken in different periods were selected from different patients of the clinical department of Russian Cardiology Research Complex (see. Table 1).

The content of total triglycerides (TG) and total cholesterol (Ch) were measured in serum by enzymatic colorimetric techniques (GPO-PAP) Triglycerides -7743 Novo, with the liquid form of Miura, 250, and (CHOD-PAP) Cholesterol-7751 Novo and Miura, 250, respectively. The individual FA triglycerides content were quantifocated with extraction, separation and methylation of FA with the use of HPLC DIONEX Ultimate 3000 using

* Corresponding author: kalinin@isan.troitsk.ru

Reprosil-Pur column C18-AQ 3 micron filler 150 × 4.6 mm. The mass spectrometer ABSCIEX QTRAP 5500 was used for registration of positively charged ions.

Table 1. Parameters of calibration sets of serum samples.

Set number (sample quantity)	Analyt, unit	Min.	Max.	Average	SD
1 (27)	ТГ, mmol/l	0.33	8.83	2.3	1.3
	X, mmol/l	0.07	7.12	3.8	1.4
2 (30)	ТГ, mmol/l	0.48	5.88	1.85	1.2
	X, mmol/l	3.0	7.16	4.51	1.17
	C16:0, μg/l	480	5880	1123	712
	C18:1, μg/l	320	5544	839	201
	C18:2, μg/l	442	6421	1021	266
3 (42)	C16:0, μg/l	305	5277	961	797
	C18:1, μg/l	253	5544	878	821
	C18:2, μg/l	419	6421	1119	949
	C20:4, μg/l	96	467	228	98

NIR spectrometry of FA triglyceride content in the serum is difficult because of dominating water absorption. Earlear, we found it possible to determine the individual FA in fat-oil mixtures with a small (6%) level of water using the overtones and combinations of fundamental absorption bands near wavelengths of 1.15, 1.4 and 1.7 μ, for which the last band is about 4 times greater than the first [4]. To isolate the FA absorption against the background of a significant water uptake, we used the coincidence of fat absorption peak, and the relative failure of the water absorption near 1.7 μ (Fig.1, a). The absorption spectra of serum samples in glass tube of 6.5 mm diameter were measured using MPA-FTNIR spectrometer, Bruker, Germany, at wavelength range 0.9 -1.8 μ, during 4 minutes of accumulation time (64 scans), and the sample temperature of 36 °C, and then smoothed by Savitsky-Goley algorithm and differentiated - see. Figure 1, b).

The spectra of optical density $D = \log I_0/I_t$, where I_0 - the flux across the empty cell, and I_t – the flux passed through the sample in the spectral range of 0.9 - 1.8 μ, with the values of mass fraction of TG, Ch, and C16: 0, C18: 1 and C18: 2 FA in each of the three sets of samples were accumulated using PLS regression program ISCAP [3] for building the PLS models. Efficiency of them was evaluated by "leave-one-out" cross validation technique in accordance with [5] by the values of multiple correlation coefficient r^2_{cv} (reliability) and standard deviation SE from arbitral data. Next, we have tested each model on serum samples from two other sets, which were not used for calibration.

Selected by the testing, the best calibration models for a range of 0.9 -1.8 μ have shown the value of the standard error SE 10.8 % and 9.3% for TG and Ch, respectively, see Table 2. The variation registered in spectra of serum allowed to estimate the correlation with the mass fraction of TG, Ch, and of FA triglycerides: C16: 0, C18: 1, C18: 2 and C20: 4. As seen from Table 2, the models to determine the C16:0, C18:1 and C18:2 FA had shown a clear correlation of spectra with FA content at a value of multiple regression coefficient r^2_p to within 0.82-0.97. With the spectra of 3rd set we have built a PLC models for the content of arachidonic C20: 4 FA. Check of its parameters by "leave-one-out" cross validation technique shows the values n = 8, r^2_{cv} = 0.68, SE = 15%, which indicates the correlation, but requires testing on samples that are not included in the calibration set.

Besides PLC models to predict the content of C16: 0, C18: 1 and C18: 2 FA, the models were built to estimate the parameters of PLS regression on the content of cholesterol, which may "prevent" prediction of other FA. It is known [4], the parameter of PLC regression - the vector of regression coefficients, (See Fig. 1c) allows one to select ranges of the spectra, which contain important data for determination of each FA and prevent the overlaping with the spectral ranges important to identify other components. For assessing the possible impact on the determination of oleic acid we have used the results of assignment of overtones and combinations in the NIR absorption spectra to the vibrations of C=C and C-

H bonds for oleic acid according to [6]. Thus obtained models GSDer and 2GSDer on 3-d set samples demonstrate the improved efficiency and accuracy for C18: 1 prediction.

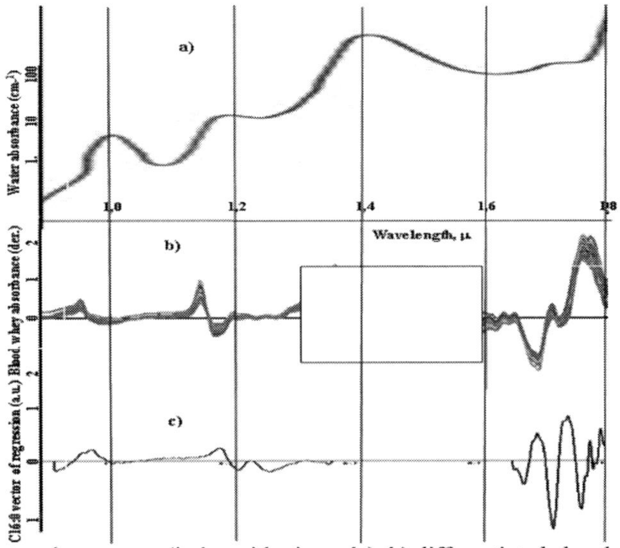

Fig. 1. Spectra: a) water absorptance (in logarithmic scale), b) differentiated absorbance of 27 serum samples, the noisy spectra – in frame, c) vector of regression coefficients for C18:1 FA.

Table 2. Parameters of prediction of TG, Ch and major FA content in independent serum samples: K is the number of calibration samples; GSDer and 2GSDer - smoothing and differentiating of the spectrum once or twice, Rb - removing spectra baseline, n - the number of latent variables; SE – standard error, %; r^2_p – multiple regression coefficient

Set № (K)	Spectra treatment	TG			Ch			C16:0			C18:1			C18:2		
		n	r^2_p	SE	n	r^2_p	SE	n	r^2_p	SE	n	r^2_p	SE	n	r^2_p	SE
1(27)	GSDer	9	0.97	11.0	4	0.86	10.9									
2(30)	GSDer	10	0.83	12.6	11	0.79	11.3	9	0.86	17	7	0.84	26	8	0.83	17
	2GSDer	9	0.96	10.8	4	0.88	9.3	8	0.87	18	7	0.86	26	8	0.85	18
3(42)	Rb	9	0.80	12.5	6	0.82	13.1	6	0.91	15	7	0.91	22	7	0.93	15
	GSDer	7	0.86	12.3	4	0.88	12.2	2	0.93	12	2	0.91	22	4	0.93	14
	2GSDer	7	0.91	12.4	5	0.87	12.2	2	0.92	13	2	0.91	24	2	0.91	18

In connection with the new concepts of the pathogenesis of atherosclerosis [7] the special diagnostic importance has the ratio of the FA C16:0 / C18:1 and C16:0 / C18:2 in the blood plasma. In the present work it was shown that these ratio C16:0/C18:1 can be quickly evaluated using NIR spectrometer in the wavelength range 0.9-1.8 μ. The prospect in this problem is to test the applicability of portable NIR spectrometer based on a linier InGaAs array photosensor of named range.

The work supported by the Russian Foundation for Basic Research, gr. № 16-02-00277.

References

1. V.N. Titov et al., Klinicheskaya Laboratornaya Diagnostika **7**, 3 (2012)
2. http://umm.edu/health/medical/altmed/supplement/omega3-fatty-acids
3. A. Kalinin et al., J. Near Infrared Spectrosc. **16**, 343 (2008)
4. A.V. Kalinin et al., Klinicheskaya Laboratornaya Diagnostika **11**, 13 (2015)
5. S. Wold, M. Sjostrom, L. Eriksson, Chem. Intelligent. Lab. Syst. **58**, 109 (2001)
6. F. Westad, A. Schmidt, M. Kermit, J. Near Infrared Spectrosc. **16**, 265 (2008)
7. V.N. Titov, Pathogenesis **11**, 18 (2013)

DOI: 10.1051/epjconf/201713203018

Electroluminescence and photosensitivity spectra of organic diode structures based on zinc complexes

M.G. Kaplunov*, S.S. Krasnikova, S.L. Nikitenko, and I.K. Yakushchenko

Institute of Problems of Chemical Physics RAS, 142432 Chernogolovka, Russia

Abstract. Devices based on zinc complexes with sulphanylaminosubstituted ligands are characterized by dual function – electroluminescence (EL) and photosensitivity. Both EL and photosensitivity are associated with the formation of exciplexes.

The zinc complexes with sulfanilaminosubstituted ligands are shown to be the efficient electroluminescent materials [1]. Light-emitting diodes (OLEDs) based on these complexes are characterized by long-wavelength emission bands that are associated with the formation of exiplexes (excited complexes) between the molecules of the emitting material (zinc complex) and hole-transport material included in the structure of the OLED [1,2]. At the same time, the formation of exciplex can result in photosensitivity of the diode structure, since exciplex is a complex with charge transfer and may arise under photoexcitation. Further charge separation under the action of an external electric field may lead to the appearance of photocurrent. Here we study photosensitivity properties of the diode structures described in ref. [1,2] as OLEDs. Figure 1 show the examples of the electroluminescence (EL) spectra and photocurrent curves (under illumination AM1.5) for the same diode structures of this type.

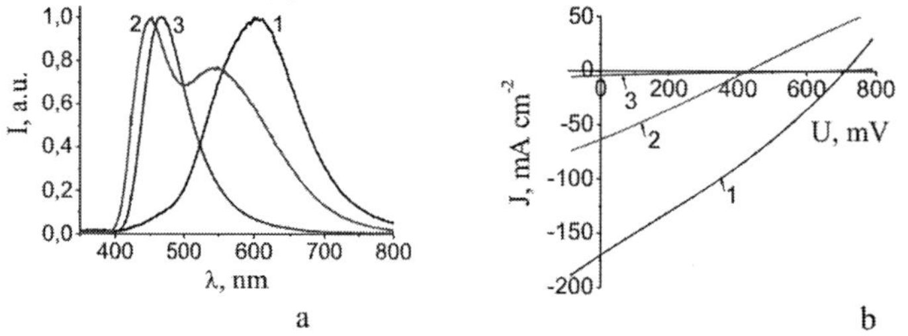

Fig. 1. The EL spectra (a) and the photocurrent curves (b) for the diode structures: (1) ITO/PTA/Zn(TSA-BTZ)$_2$/Al:Ca, (2) ITO/PTA/NPD/Zn(TSA-BTZ)$_2$/Al:Ca and (3) ITO/PTA/CBP/Zn(TSA-BTZ)$_2$/Al:Ca.

The efficiency of the exciplex formation depends on the material of the hole-transporting layer contacting with the zinc complex. In the EL spectra, this dependence manifest itself in the ratio of the intensities of the intrinsic radiation band of the zinc complex in the region of 450 nm and the exciplex radiation band in the region of 550 –

* Corresponding author: kaplunov@icp.ac.ru

600 nm (Figure 1a). The efficiency of the exciplex formation is reduced in the series of the hole-transporting msterials PTA-NPD-CBP. Only exciplex band at 600 nm, both the intrinsic band at 450 nm and the exciplex band at 545 nm and only intrinsic band at 465 nm are observed correspondingly for PTA, NPD and CBP contacting with the zinc complex. In case of photosensitivity curves, the photocurrent is reduced in the same series PTA-NPD-CBP in accordance with the efficiency of the exciplex formation (Figure 1b). This fact confirms the exciplex nature of the appearance of the photocurrent.

The study of the spectral dependence of photosensitivity showed that the maximum photocurrent is observed under illumination by light in the region of 350-450 nm, which corresponds to the spectral region of absorption of the materials forming the diode structure. It is of interest to find out which components of the device are responsible for the primary absorption of a light quantum in the formation of the exciplex. For this purpose, we measured the photoluminescence (PL) excitation spectra of thin films containing Zn(TSA-BTZ)$_2$, PTA and their mixture 1:1 by weight, and their PL and absorption spectra (figure 2).

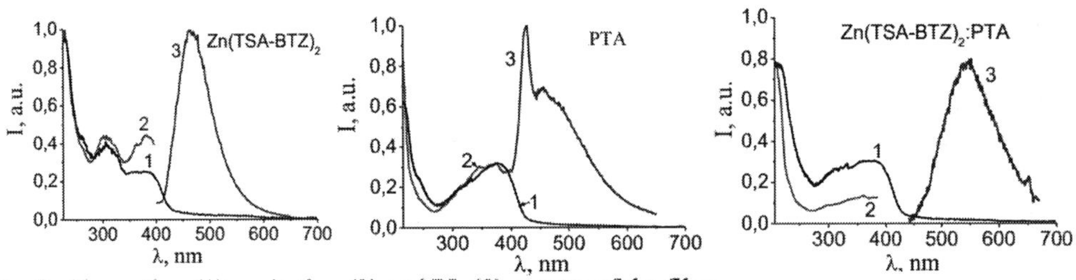

Fig. 2. Absorption (1) excitation (2) and PL (3) spectra of the films.

In the PL spectra of Zn(TSA-BTZ)$_2$ and PTA, bands of the intrinsic PL of these substances are observed respectively at 470 nm and 425-460 nm, while the exciplex band at 550 nm is observed in the PL spectrum of the mixture. In the absorption and the excitation spectra of the film of pure Zn(TSA-BTZ)$_2$ complex, a band around 300 nm is observed. Such band is absent in the absorption and the excitation spectra of PTA. Herewith the absorption band around 300 nm is present in the absorption spectrum of the mixture, which is obviously due to the absorption of Zn(TSA-BTZ)$_2$ molecules. However, this band is absent in the excitation spectrum of the mixture. This is due to the exciplex nature of the fluorescence of the mixture. One may suppose that in the case of PL of the mixture, the primary excitation occurs on the PTA molecule with subsequent electron transfer to an excited level of the Zn(TSA-BTZ)$_2$ molecule and then the radiative recombination with the hole on the main level of PTA. In the case of photocurrent in the device based on Zn(TSA-BTZ)$_2$ and PTA, the primary excitation obviously occurs also on the molecule of the hole transport layer PTA with subsequent separation and transport of charge carriers to the electrodes.

Thus we show that the diode structures described in ref. [1,2] as OLEDs are also characterisrd by photosensitivity properties. which is associated with the formation of exciplexes. The primary process in the photosensitivity is the photoexcitation of the molecules of the hole-transporting layer.

This work was supported by Russian Foundation for Basic Research (16-33-01091).

References

1. M.G. Kaplunov, S.S. Krasnikova, S.L. Nikitenko, I.K. Yakushchenko, Nanotechnologies in Russia 7, 415 (2012)
2. M.G. Kaplunov, S.S. Krasnikova, S.L. Nikitenko, N.L. Sermakasheva, I.K. Yakushchenko, Nanoscale Research Letters 7, 206 (2012)

EPJ Web of Conferences **132**, 03019 (2017) DOI: 10.1051/epjconf/201713203019
SPECTROSCOPY.SU 2016

Spectrum of five times ionized tin (Sn VI) in the vacuum UV region

R.R. Kildiyarova*, E.Ya. Kononov, and A.N. Ryabtsev

Institute for Spectroscopy RAS, 108840 Troitsk, Moscow, Russia

Abstract. The spectrum of tin excited in a vacuum spark was recorded in the region 500 – 1500 Å on the 6.65 m normal incidence spectrograph. 110 lines of the $4d^85p$ – $4d^85d$ transitions in the Sn^{5+} ion in the region 680 – 850 Å have been identified and 52 energy levels of the $4d^85d$ configuration have been established.

Five times ionized tin, Sn VI, has the ground configuration $4d^9$. The resonance $4d^9$ - $4d^85p$ transitions and the transitions between the first excited configurations $4d^85s$ - $4d^85p$ have been studied previously [1,2]. After that, in our work [3] the transitions from higher excited configurations $4d^9$ - $4d^8(4f+6p)$ were added. As a result, the levels of the excited $4d^84f$, $4d^85s$, $4d^85p$ and $4d^86p$ were found. The only $4d^85d$ configuration of the n=5 sub shell was still unknown. In the present work the $4d^85p$ - $4d^85d$ transitions were studied and the levels of the $4d^85d$ configuration were located. This transition array is promising for a development of a radiation source in the ~ 800 Å region, which could be used for a molecular hydrogen dissociation in a problem of the optics cleaning in the lithography machines working with a tin radiation source on the 135 Å wavelength.

The spectra of tin were photographed on Ilford Q plates on a 6.65 m normal incidence spectrograph equipped with a 1200 l/mm grating. Excitation of the spectra was accomplished in a three-electrode vacuum spark with the parameters of discharge circuit: C = 10 μF, U = 3÷5 kV, L = 100÷500 нH. The discharge conditions were adjusted to have the high intensity of known Sn VI lines. Atomic calculations for the Sn^{5+} ion were performed using the Cowan code. In the first approximation the Hartree-Fock energy parameters of the $4d^85d$ configuration were scaled by the factors derived from the known $4d^95d$ configuration in Sn V. As new information on the $4d^85d$ levels becomes available the parameters were semi-empirically adjusted by a fitting to known energy levels resulting in more accurate spectrum prediction.

As a result, 110 lines of the $4d^85p$ – $4d^85d$ transitions in the region 680 – 850 Å were identified and 52 energy levels of the $4d^85d$ configuration were found in Sn^{5+}. After final fitting of the calculated to experimental level energies the wave functions in the intermediate coupling were obtained and used for the calculations of the transition probabilities. A comparison of the calculated transition probabilities (gA-values) with the observed line intensities of the $4d^85p$ – $4d^85d$ transitions is shown in Fig. 1.

* Corresponding author: rimma@isan.troitsk.ru

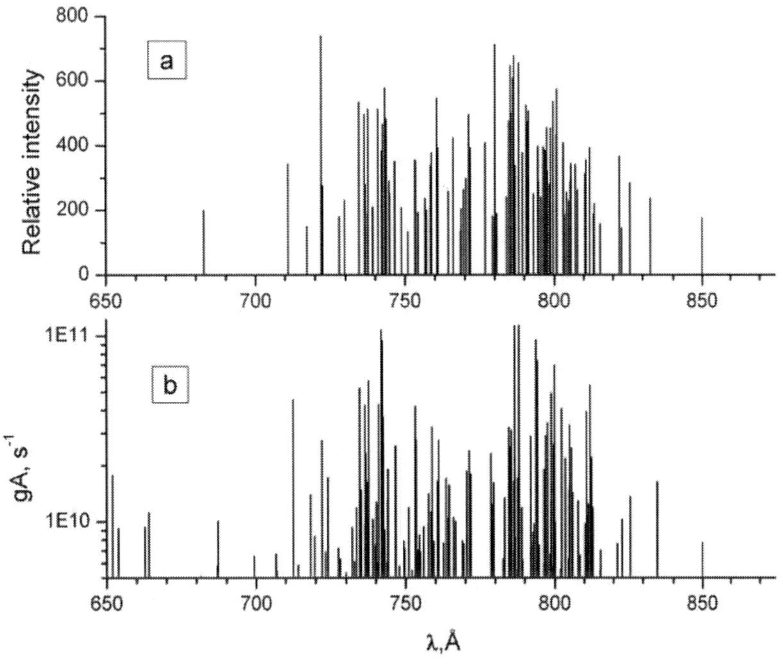

Fig. 1. Observed (a) and calculated (b) $4d^85p$ - $4d^85d$ transitions in the Sn^{5+} spectrum.

The agreement of the calculated transition probabilities with the line intensities is quite good taking into account nonlinearity of photographic response and possible blending of some lines. Thus, these atomic data can be helpful for plasma modelling in a development of the radiation source for ~ 800 Å region.

The work was supported by the Russian Foundation for Basic Research, grant № 16-02-00753a.

References

1. Y.N. Joshi, Th.A.M. Van Kleef, Can. J. Phys. **55**, 714 (1977)
2. R.P. Srivastava, Y.N. Joshi, Th.A.M. Van Kleef, Can. J. Phys. **55**, 1936 (1977)
3. S.S. Churilov, R.R. Kildiyarova, A.N. Ryabtsev, A.E. Kramida, Y.N. Joshi, Phys. Scr. **50**, 463 (1994)

EPJ Web of Conferences **132**, 03020 (2017) DOI: 10.1051/epjconf/201713203020
SPECTROSCOPY.SU 2016

Plasma heating and acceleration in current sheets formed in discharges in argon

N.P. Kyrie*

Prokhorov General Physics Institute RAS, 119991 Moscow, Russia

Abstract. According to present notion, flares on the sun and other stars, substorms in magnetospheres of Earth and other planets, and disruptive instabilities in tokamak plasma are connected to development of current sheets in magnetized plasma. Therefore, current sheet dynamics and magnetic reconnection processes were studied actively during the last several decades. This paper presents the results of experimental studies of plasma heating and acceleration in current sheets formed in discharges in argon. The temperature and energy of directed motion of argon ions of different degrees of ionization were measured by spectroscopic methods. It was found that Ar II, Ar III and Ar IV ions are localized in different regions of the sheet. It was shown that Ampere forces applied to the sheet can accelerate the argon ions to observed energies.

The paper is concerned with measurements of the temperature and energy of directed motions of ions in laboratory current sheets formed in two-dimensional three-dimensional configurations with a singular X line in argon. According to theoretical notions [1], both thermal processes and plasma flows can play a decisive role in passing of a current sheet from the metastable to impulsive phase of magnetic reconnection [2].
The studies were conducted by spectroscopic methods to measure profiles of the Ar II 480.6 nm, Ar III 379.5 nm, Ar IV 280.9 nm spectral lines broadening due to Doppler Effect. The plasma radiation was collected and analyzed with the help of an optical scheme. The measurements were made in two mutually perpendicular directions, i.e. along the current (the z-axis) and along the width that is a larger size of a current sheet (the x-axis) [3-5]. The spectral line profiles were measured from the line intensity at the exit slit of a monochromator by scanning the spectrum with the help of a "Nanogate 1-UF"digital image intensifier [6]. This allowed us to measure the profile in one shot of the device.

For the current sheets formed in two-dimensional magnetic configurations (p = 28 mTorr, I_z = 45 kA, h = 0.5 kG/cm, B_z = 0), it was found that Ar II, Ar III and Ar IV ions of argon are localized in different regions of the current sheet, depending on their temperature. The Ar II ions are located in relatively cold plasma regions at the periphery with a maximum temperature T_i = (60 ± 15) eV, whereas the Ar III and Ar IV ions are located in hotter central regions of the current sheet with a maximum temperature T_i = (100 – 140) eV.

* Corresponding author: kyrie@fpl.gpi.ru

For all spectral lines of argon ions, the FWHM (full width half maximum) values measured in the x-direction are great in comparison with the same lines measured in the z-direction. We interpreted this result to mean that the Ar II, Ar III, Ar IV ions participate not only in thermal, but also in directed motion. Such directed motion along the sheet surface (the x-axis) is interpreted as an effect of Ampere forces. It is found that all argon ions can equally gain excess velocity of directed motion $V_x \approx (2.5 \pm 0.3)\times10^6$ cm/s which exceeds the thermal velocity of the Ar II ion, but is less than (or equal to) the thermal velocity of the Ar III and Ar IV ions.

From magnetic measurements conducted under the same experimental conditions, it follows that the accelerated plasma flows with time intersect the regions of strong transverse magnetic field at edges of the current sheet and generate reverse currents in these regions [7,8]. Estimates show that the Ampere forces in a current sheet can accelerate argon ions up to the observed energies [9,10].

For the current sheets formed in the three-dimensional magnetic configurations (p = 28 mTorr, I_z = 45 kA, h = 0.64 kG/cm, B_z = 0 – 4.35 kG), we investigated how the temperature and energy of directed motion of the Ar II ions depends on the induction of guiding magnetic field B_z in the range = 0 – 4.35 kG. It was found that the temperature of the Ar II ions gradually decreases with increasing B_z from 0 to 1.45 kG: T_i = 31±3 eV is reduced to T_i = 23±2 eV. At the same time, the energy of directed motion of the Ar II ions increases but somewhat from W_x = 50±13 eV to W_x = 56±16 eV (within the limits of large experimental error). Above the point B_z = 1.45 kG, on the contrary, the temperature of Ar II ions increases and eventually reaches its maximum value T_i = 44±4 eV at B_z = 4.35 kG; the energy of directed motion of the Ar II ions is reduced to W_x = 31±14 eV.

This work pursued the Program of Russian Academy of Sciences, under state registration number No. 01200953488; the author was supported in part by RFBR, under No.15-02-03644.

References

1. S.I. Syrovatskii, Annu. Rev. Astron. Astrophys. **19**, 163 (1981)
2. E.R. Priest and T. Forbes, *Magnetic Reconnection: MHD Theory and Applications* (Cambridge Univ. Press, New York, 2000)
3. N.P. Kyrie, V.S. Markov, A.G. Frank, Plasma Phys. Rep. **36**, 357 (2010)
4. N.P. Kyrie, V.S. Markov, A.G. Frank, JETP Lett. **95**, 14 (2012)
5. N.P. Kyrie, A.G. Frank, Plasma Phys. Rep. **38**, 960 (2012)
6. http://www.nanoscan.ru
7. A.G. Frank, S. N. Satunin, Plasma Phys. Rep. **37**, 829 (2011)
8. A.G. Frank, N.P. Kyrie, S.N. Satunin, Physics of Plasmas. **18**, 111209 (2011)
9. N.P. Kyrie, V. S. Markov, A. G. Frank, D.G. Vasilkov, E.V. Voronova, Plasma Phys. Rep. **42**, 549 (2016)
10. A.G. Frank, N.P. Kyrie, Uspekhi Prikladnoy Fiziki **3**, 454 (2015) (in Russian)

DOI: 10.1051/epjconf/201713203021

Raman scattering on electronic levels in rare-earth iron borates $R\mathrm{Fe_3(BO_3)_4}$

S.A. Klimin*, and M.N. Popova

Institute for Spectroscopy RAS, 108840 Troitsk, Moscow, Russia

Abstract. We have performed a Raman scattering study for single crystals of the rare-earth (RE) iron borates $R\mathrm{Fe_3(BO_3)_4}$, R = Nd, Tb, Pr, in the range of temperatures 3-300 K. Electronic Raman scattering on the 4f levels of RE ions was registered, which enabled us to specify the schemes of crystal-field levels within ground multiplets of the RE ions. A splitting of the Kramers doublets in electronic Raman spectra of the neodymium iron borate testifies a magnetic ordering of the compound. Raman scattering from a coupled mode of the spin oscillations of the Fe and Nd subsystems was observed in $\mathrm{NdFe_3(BO_3)_4}$.

The rare-earth (RE) iron borates $R\mathrm{Fe_3(BO_3)_4}$ are widely studied by different methods due to their multiferroicity. They are interesting both as promising compounds and as model systems for studying mechanism of interactions between different degrees of freedom such as charge, magnetic, lattice, and electronic ones [1-4]. All the iron borates experience phase transition into antiferromagnetic phase at low temperatures in the range 32-39 K. Precise information on the energies of crystal-field (CF) levels for the ground multiplet of the RE ion is needed to interpret reliably magnetic, magnetoelectric, and thermal properties of these compounds, as well as the effects of electron-phonon and electron-magnon interactions. Optical IR transmission spectroscopy gives limited information on the CF structure of the ground multiplet, as the population of excited levels is small, while Raman spectroscopy does not suffer from this restriction. In this work, we study Raman scattering on electronic levels of RE iron borates with R = Nd, Tb, Pr.

Raman spectra in a wide temperature range (3-300 K) were measured using Jobin-Yvon T64000 spectrometer. At low temperatures, in the Raman spectra registered, in addition to phonon peaks, spectral lines are present related to two-magnon as well as to electronic scattering. We have used several excitation wavelengths (532, 514.5 и 488 nm) to separate a scattered signal from a luminescent one.

Several lines found in the Raman spectra of iron borates were attributed to the electronic scattering on the crystal-field levels of RE ions, namely, 201 см$^{-1}$ for $\mathrm{Tb^{3+}}$, 197 см$^{-1}$ for $\mathrm{Pr^{3+}}$, and 140 and 220 см$^{-1}$ for $\mathrm{Nd^{3+}}$. At the temperatures lower than the temperature of an antiferromagnetic ordering of $\mathrm{NdFe_3(BO_3)_4}$, T_N = 33 K, a new peak arises in the polarization geometry *zx* due to the scattering on the upper component of the split ground Kramers doublet of the $\mathrm{Nd^{3+}}$ ion (Fig. 1). This line moves to higher energy values with cooling and reaches 11 см$^{-1}$ at $T \approx 5$ K. This value differs from the value Δ_0 = 8.8 см$^{-1}$ for

* Corresponding author: klimin@isan.troitsk.ru

the splitting of the ground Kramers doublet of the Nd^{3+} ion found by optical IR-study of $NdFe_3(BO_3)_4$ [5], while it is in a good agreement with the data obtained by submillimeter spectroscopy [6]. Similarly to the case of submillimeter radiation incident on the crystal, in the process of Raman scattering on 4f electronic CF levels, a new 4f-excitation appears in the crystal. This excitation interacts with close in energy antiferromagnetic mode of Fe-magnetic subsystem, which leads to the formation of coupled modes for the vibrations of the spins of Fe- and Nd-subsystems. Renormalization of frequencies [6,7] is responsible for the observed phenomenon.

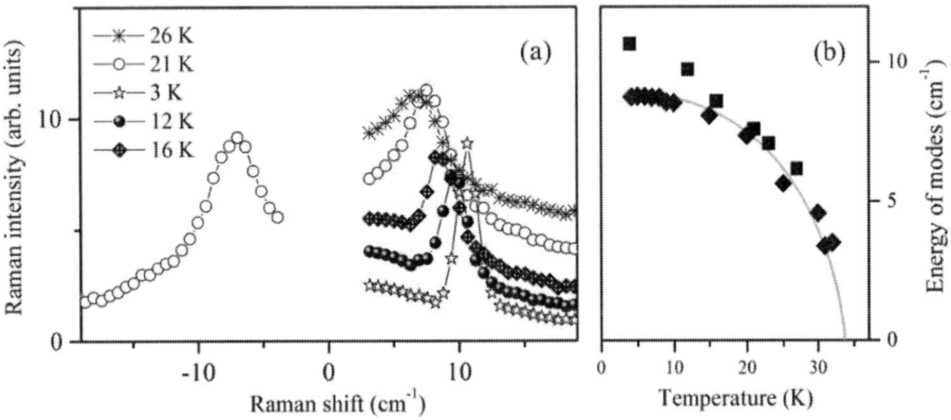

Fig. 1 (a) Raman spectra at different temperatures lower than T_N. Anti-Stokes component is shown confirming the origin of a peak presented. (b) Data on the splitting of ground Kramers doublet of Nd^{3+} ion (rhombs) [5] and the energy of the Raman mode (squares, this study) as a functions of temperature. Gray curve is guided to the eye.

Support by Russian Science Foundation under Grant No. 14-12-01033 is acknowledged.

References

1. A.M. Kadomtseva, Yu.F. Popov, G.P. Vorob'ev, A.P. Pyatakov, S.S. Krotov, K.I. Kamilov, V.Yu. Ivanov, A.A. Mukhin, A.K. Zvezdin, A.M. Kuz'menko, L.N. Bezmaternykh, I.A. Gudim, V.L. Temerov, Low Temp. Phys. **36**, 511 (2010)
2. M.N. Popova, K.N. Boldyrev, S.A. Klimin, T.N. Stanislavchuk, A.A. Sirenko, L.N. Bezmaternykh, J. Magn. Magn. Mater. **383**, 250 (2015)
3. S.A. Klimin, A.B. Kuzmenko, M.A. Kashchenko, M.N. Popova, Phys. Rev. B **93**, 054304 (2016)
4. K.N. Boldyrev, T.N. Stanislavchuk, S.A. Klimin, M.N. Popova, L.N. Bezmaternykh, Physics Letters A **376**, 2562 (2012)
5. M.N. Popova, E.P. Chukalina, T.N. Stanislavchuk, B.Z. Malkin, A.R. Zakirov, E. Antic-Fidancev, E.A. Popova, L.N. Bezmaternykh, V.L. Temerov, Phys. Rev. B **75**, 224435 (2007)
6. A.M. Kuz'menko, A.A. Mukhin, A.M. Kadomtseva, V.Yu. Ivanov, L.N. Bezmaternykh, JETP **113**, 113 (2011)
7. A.V. Peschanskii, V.I. Fomin, I.A. Gudim, Low Temp. Phys. **42**, 475 (2016)

Effect of magnetic ordering of Dy_2BaNiO_5 on the crystal-field levels of dysprosium: optical spectroscopy of f-f transitions

A.S. Galkin[1,2], and S.A. Klimin[1,*]

[1]Institute for Spectroscopy RAS, 108840 Troitsk, Moscow, Russia
[2]Technological Institute for Superhard and Novel Carbon Materials, 108840, Troitsk, Moscow, Russia

Abstract. Optical transmission spectroscopy study of the Haldane magnet Dy_2BaNiO_5 was performed in the region of f-f transitions of the Dy^{3+} ion in a wide range of temperatures (5-300 K). At temperatures lower than T_N (59 K), Kramers doublets of the rare-earth ion split. Spectroscopic data obtained were used to calculate the Schottky-type anomaly in the temperature dependence of the magnetic susceptibility of Dy_2BaNiO_5 and to model the experimental data available in literature. Anomalous behavior of crystal-field energies of the Dy^{3+} ion was attributed to the magnetoelectric interactions.

Dysprosium nickelate Dy_2BaNiO_5 belongs to the family of rare-earth (RE) chain nickelates R_2BaNiO_5 (R=RE), well-known as model compounds for studying Haldane magnetism [1,2]. In Dy_2BaNiO_5, antiferromagnetic (AFM) ordering arises at $T_N = 59$ K [3,4] due to dysprosium-mediated interaction between nickel S=1 chains. Temperature dependence of the magnetic susceptibility $\chi(T)$ of Dy_2BaNiO_5 has a broad maximum at $T_m = 45$ K [5], attributed to a dysprosium contribution. Shottky-type anomaly in the $\chi(T)$ dependence can be calculated directly from spectroscopic data on the RE ground-doublet splitiing [6,7], however, such data are not available in the literature. According to recent studies, Dy_2BaNiO_5 demonstrates multiferroic behavior [8], spontaneous electric polarization **P** emerges at $T<T_N$. In this work, we study the temperature behavior of crystal-field (CF) levels (Kramers doublets) of the Dy^{3+} ion by means of optical spectroscopy, with the aim to extract information on low-energy CF-excitations of the dysprosium ion and to calculate the dysprosium contribution into the magnetic susceptibility of Dy_2BaNiO_5. Another goal was to look for spectroscopic effects related to the appearance of **P** in the AFM phase.

Optical transmittance spectra of polycrystalline Dy_2BaNiO_5 were measured in a wide spectral range (3000 – 15000 cm^{-1}) comprising several multiplets of the Dy^{3+} ion, with the use of a BRUKER IFS125HR Fourier spectrometer. To get low temperatures, a closed-cycle optical cryostat CRYOMECH PT403 was used. The analysis of temperature-dependent transmission spectra enabled us to find the following CF energies (in cm^{-1}) for different multiplets: $^6H_{15/2}$ (0, 150, 275), $^6H_{13/2}$ (3565, 3670, 3690, 3770, 3810, 3860, 3910), $^6H_{11/2}$ (5978, 6026, 6063 6070, 6109), $^6H_{9/2}$+$^6F_{11/2}$ (7755, 7788, 7824, 7923, 7946, 7962,

* Corresponding author: klimin@isan.troitsk.ru

7985, 8033, 8135),), $^6H_{7/2}$+$^6F_{9/2}$ (9138, 9173, 9220, 9274, 9343, 9366, 9422, 9518), $^6F_{7/2}$ (11095, 11170, 111925, 11258), $^6F_{5/2}$ (12505, 12525, 12630), $^6F_{3/2}$+$^6F_{1/2}$ (13353). In particular, the ground Kramers doublet is separated by a gap from the excited states. Thus, a model of an isolated ground doublet [6,7] can be used for describing thermodynamic properties of Dy_2BaNiO_5.

The observed splitting of Kramers doublets (Fig. 1) confirms a magnetic ordering of the crystal at T_N = 59 K. For modeling of the dysprosium contribution to the magnetic susceptibility $\chi(T)$ of Dy_2BaNiO_5 we have utilized experimentally obtained temperature-dependent splitting of the ground Kramers doublet $\Delta_0(T)$, which reaches the value of 39 cm^{-1} at 5 K. The results of modeling lead us to a conclusion that the low-temperature magnetic susceptibility of Dy_2BaNiO_5 is determined by the dysprosium contribution, while the one-dimensional contribution of the nickel chains is much smaller.

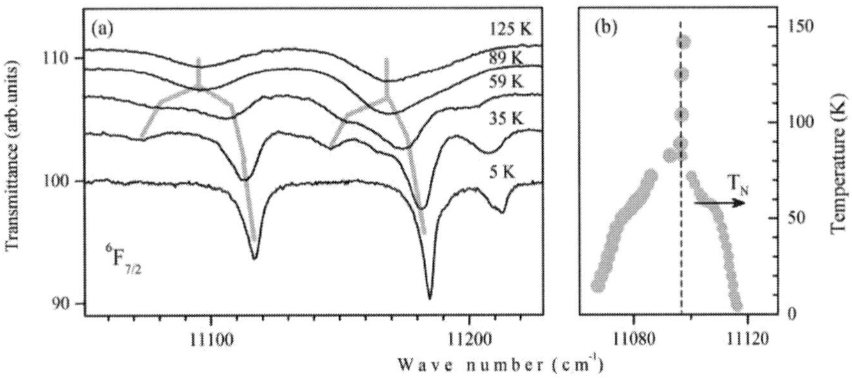

Fig. 1. (a) Transmission spectra in the region of $^6F_{7/2}$ multiplet of Dy^{3+} ion in Dy_2BaNiO_5 at different temperatures. Grey lines show schematically line splittings. (b) Positions of line 11096 cm^{-1} and its split components as a function of temperature. Vertical dashed line is drawn to underline asymmetrical character of splitting.

Unusual behavior of the dysprosium CF energies in Dy_2BaNiO was detected in the temperature range lower than T_N. All CF levels experience a shift (up to 10 cm^{-1}) from their positions at T_N (see, e.g., Fig. 1). Such behavior can be explained as being due to the spontaneous electric polarization that appears in the temperature range discussed [8]. Redistribution of the charge density influences the crystal field acting on dysprosium and leads to a renormalization of CF self-energies.

References

1. J. Darriet, L.P. Regnault, Solid State Commun. **86**, 409 (1993)
2. S.A. Klimin, A.B. Kuzmenko, M.N. Popova, B.Z. Malkin, I.V. Telegina, Phys. Rev. B **82**, 174425 (2010)
3. E. García-Matres, J.L. Martínez, J. Rodríguez-Carvajal, Eur. Phys. J. B **24**, 59 (2001)
4. S.A. Klimin, A.S. Galkin, M.N. Popova, Phys. Lett. A **376**, 1861 (2010)
5. E. García-Matres, J.L. García-Muũnoz, J.L. Martínez, J. Rodríguez-Carvajal, J. Magn. Magn. Mater. **149**, 363 (1995)
6. E.A. Popova, S.A. Klimin, M.N. Popova, R. Klingeler, N. Tristan, B. Buchner, A.N. Vasiliev, J. Magn. Magn. Mater. **331**, 133 (2013)
7. S.A. Klimin, A.S. Galkin, M.N. Popova, J. Alloys Comp. **625**, 193 (2015)
8. Kiran Singh, Tathamay Basu, S. Chowki, N. Mahapotra, K. Iyer Kartik, P. L. Paulose, E.V. Sampathkumaran, Phys. Rev. B **88**, 094438 (2013)

Exited configurations in the spectrum of five-times ionized indium (In VI)

E.Ya. Kononov[1,*], R.R. Kildiyarova[1], A.N. Ryabtsev[1], A. Tauheed[2], Swapnil[2], Hala[2], and A. Saxena[2]

[1] Institute for Spectroscopy RAS, 108840 Troitsk, Moscow, Russia
[2] Aligarh Muslim University, 202002 Aligarh, India

Abstract. An Indium spectrum was obtained in the region 160 – 230 Å using the grazing-incidence spectrograph (incidence angle 85°, grating radius 3 m, 3600 lines/mm) and a vacuum spark at peak currents 12 – 35 kA. A preliminary identification of the $4d^8$ - $4d^74f$ and $4d^8$ - $4d^75f$ transitions in In VI have been done.

The most efficient radiation source for photo lithography on wavelength 135 Å is a high-temperature plasma with multiply ionized atoms of tin. It is well established that radiative transitions of Sn IX – Sn XIV are present in band 135 Å ±1% of available mirrors. Developers of the radiation source seek for increasing its efficiency by creating such specific plasma conditions which maximize intensity in the required band relatively that out of band. The most important is the long-wave region where radiation of lower ionization stages is present. Information on such spectra is either scarce or insufficiently reliable, requiring verification by comparison with isoelectronic spectra of neighbouring chemical elements, in the first turn, with indium spectra. Spectra of indium ions are interesting themselves for creation of a radiation source since a mixture of tin and indium is suggested for gain enhancing. Below is a progress report on the investigation of the five times ionized indium.

The spectrum of the five times ionized indium has been studied earlier elsewhere [1, 2]. The ground state configuration of In VI is $4p^64d^8$. The transitions $4d^8$ - $4d^75p$ and $4d^75s$ – $4d^75p$ were investigated. The energy levels of the all three relevant configuration were established. Our project presumes a study of the other class of transitions, namely $4p^64d^8$ - ($4p^54d^9$ + $4p^64d^74f$ + $4p^64d^75f$ + $4p^64d^76p$). Exactly this class of transitions in ions of higher ionization stages having large transition probabilities creates in tin in band radiation.

The indium spectrum in the region 160 – 230 Å (where the mentioned transitions are expected according to calculations) was obtained using the grazing-incidence spectrograph (incidence angle 85°, grating radius 3 m, 3600 lines/mm). Spectra were excited in a vacuum spark with varied voltage (1 – 5 kV), capacity (12 or 150 μF) and inductance (100 – 930 nH), and as a result the peak current (12 – 35 kA) for differentiation of spectral lines on ionization stages. Spectra were recorded on FUJI Imaging Plates. The latter were digitized on the Typhoon FLA 9500 scanner. The high linearity of imaging plates compared with

* Corresponding author: kononov@isan.troitsk.ru

photographic plates allowed more reliable measurements of line intensity changes when spark discharge conditions were varied.

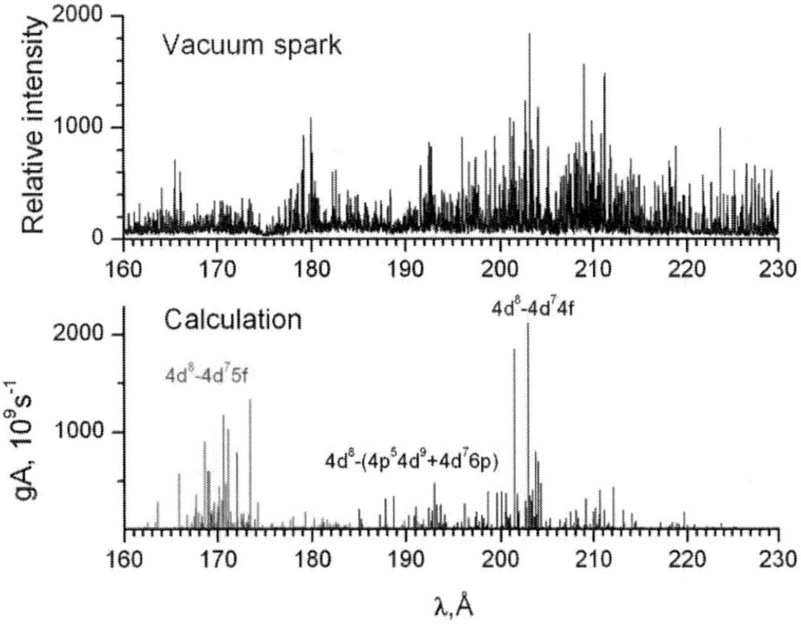

Fig. 1. Vacuum spark spectrum of indium and calculated transition probabilities in In VI.

The upper part of Fig. 1 shows an indium spectrum at peak current 12 kA. The lower part presents calculated transition probabilities (gA) in In VI. They were calculated by the Hartree-Fock method with relativistic corrections and semi-empirical scaling of energy parameters. The scaling factors were derived from the Sn VII isoelectronic spectrum.

More than 1600 spectral lines were measured on the spectrograms, among them more than 200 were attributed to In VI due to variations of their intensities with discharge conditions in the spark. A preliminary identification of the $4d^8$ - $4d^74f$ and $4d^8$ – $4d^75f$ transitions have been made.

The work was supported by the Russian Foundation for Basic Research, project № 16-52-45036 IND_a.

References

1. V.S. Kushawaha, Y.N. Joshi, JOSA **66**, 630 (1976)
2. Th.A.M. Van Kleef, Y.N. Joshi, JOSA **72**, 1348 (1982)

3D imaging of semiconductor colloid nanocrystals: on the way to nanodiagnostics of track membranes

S.I. Kulyk[1,2,*], I.Y. Eremchev[1], A.A. Gorshelev[1], A.V. Naumov[1], D.L. Zagorsky[3], S.P. Kotova[4], V.G. Volostnikov[4], and E.N. Vorontsov[4]

[1]Institute for Spectroscopy RAS, 108840 Troitsk, Moscow, Russia
[2]Moscow Institute of Physics and Technology, 141700 Dolgoprudny, Russia
[3]Schubnikov Institute of crystallography, 119333 Moscow, Russia
[4]Samara branch of the Physical institute of RAS, 443011 Samara, Russia

Abstract. The work concerns the feasibility of 3D optical diagnostic of porous media with subdifraction spatial resolution via epi-luminescence microscopy of single semiconductor colloid nanocrystals (quantum dots, QD) CdSe/ZnS used as emitting labels/nanoprobes. The nanoprecise reconstruction of axial coordinate is provided by double helix technique of point spread function transformation (DH-PSF). The results of QD localization in polycarbonate track membrane (TM) is presented.

Track etched membranes are widely used in industry (high efficiency water filters) and science (metal nanostructure growing, optical analysis of DNA) [1]. However, the study of physical and chemical properties of such porous media is entangled due to the small sizes of pores (hundreds of nanometers and less). Thus methods of electronic microscopy have several restrictions, and probe microscopy (atomic-force microscopy) allows to only characterize the surface, giving no answer for the question of structure and properties of the pores' walls.

In recent years the characterization of materials with complex structure and objects at nanometer scale by methods of super resolution optical microscopy with detection of single luminescence emitter-probes has become more and more popular ([2, 3] and links inside). The potential of such methods can be fulfilled in a nano diagnostics of complex polymer structures, and track membranes also. Moreover, the photophysical (spectral) characteristics of such probes are very sensitive to local environment, that can be used in future for characterization of that environment [2-7].

Determination of all three spatial coordinates with nanometer precision is possible due to use of an idea of instrumental transformation of point spread function in double helix scheme using adaptive optics' phase transformation elements [8].

The purpose of introduced work is development of methods of 3D laser spectromicroscopy of single colloid nanocrystals (quantum dots, QD) for nanodiagnostics of track etched membranes.

* Corresponding author: sergiy.i.kulyk@gmail.com

In present paper the experimental test of original calculated phase transformation elements (masks), that allow the transformation of point spread function in double helix scheme, was conducted [9]. Also it was shown, that the accuracy of reconstruction of coordinates of quantum dots (CdSe/ZnS), inserted in liquid solution in polycarbonate track membranes, reaches several dozens of nanometers. The separation between individual images of different molecules in each diffracted limited volume of the sample is possible due to inherent QD property of photoluminescence blinking [10-16].

The important part of the work was the choice of "sample-probe" pair. As far as the accuracy of coordinate reconstruction depends on signal-to-noise ratio, the background luminescence of the sample could significantly spoil or even make impossible the detection of single emitters inside the pores. Due to weak cross-section of QD CdSe/ZnS and polycarbonate TM luminescence spectra the precision of 20-40 nanometer for 400-500 ms exposure was reached.

Also the feasibility of dynamical processes' diagnostics by introduced method was shown – for example, the visualization of filtration process with the help of markers, in real time with nanometer precision.

Authors acknowledge the support of Russian Foundation of Basic Researches (projects 14-02-00822-a and 16-29-11809-ofi_m). Authors are also grateful professor P.U. Apel (Frank Laboratory of Neutron Physics, Joint Institute for Nuclear Research, Dubna) for attested samples of polymer track etched membranes.

References

1. V.V. Beriozkin et al., Radiation Measurements **34**, 593 (2001)
2. A. Naumov, I. Eremchev, A. Gorshelev, Eur. Phys. J. D **68**, 348 (2014)
3. A.V. Naumov, Physics Uspekhi **56**, 605 (2013)
4. T. Plakhotnik, E. A. Donley, U. P. Wild, Ann. Rev. Phys. Chem. **48**, 181 (1997)
5. T. Plakhotnik, D. Walser, A. Renn, Phys. Rev. Lett. **77**, 5365 (1996)
6. T. Plakhotnik, J. Lumin. **83-84**, 221 (1999)
7. M. Orrit et al., Faraday Discussions **184**, 275 (2015)
8. S. Pavani, R. Piestun, Opt. Express **16**, 3484 (2008)
9. S.P. Kotova et al., J. Phys.: Conf. Ser. **740**, 012013 (2016)
10. M. Nirmal et al., Nature **383**, 802 (1996)
11. P. Frantsuzov, M. Kuno, B. Janko, R. A. Marcus, Nature Physics **4**, 519 (2008)
12. A.L. Efros, D.J. Nesbitt, Nature Nanotechnology **11**, 661, (2016)
13. T. Plakhotnik et al., Phys. Rev. Lett. **105**, 167402 (2010)
14. M.J. Fernee et al., J. Phys. Chem. Lett. **3**, 1716 (2012)
15. I.S. Osad'ko, I.Yu. Eremchev, A.V. Naumov, J. Phys. Chem. C **119**, 22646 (2015)
16. I.Yu. Eremchev, I.S. Osad'ko, A.V. Naumov, J. Phys. Chem. C **120**, 22004 (2016)

Intramolecular vibrational dynamics of ν_1 mode in $(CF_3)_2C{=}C{=}O$ molecule induced by resonant IR femtosecond radiation

S.V. Chekalin[1], V.O. Kompanets[1], V.B. Laptev[1,*], A.A. Makarov[1], S.V. Pigul'sky[1], E.A. Ryabov[1], G.V. Mishakov[2], A.V. Sharkov[3], D.V. Serebryakov[4], I.A. Denisov[5], and N.A. Smirnova[5]

[1]Institute for Spectroscopy RAS, 108840 Troitsk, Moscow, Russia
[2]CDP Systems Corp., 119991 Moscow, Russia
[3]Lebedev Physical Institute RAS, 119991 Moscow, Russia
[4]Institute for nuclear research RAS, 108840 Troitsk, Moscow, Russia
[5]JSC "Giredmet", 119017 Moscow, Russia,

Abstract. Intramolecular dynamics of the vibrational levels (up to v = 5) of the ν_1 mode in $(CF_3)_2CCO$ molecule, induced via multiphoton selective excitation of this mode by resonant femtosecond IR radiation, was studied. The times of intramolecular vibrational energy redistribution (*IVR*) of each vibrational level v to the remaining modes of the molecule were measured. In accordance with the theory predictions the decrease of *IVR* time with increasing v is observed. A sharp reduction of the *IVR* time (up to 1 ps) at a wavelength of 2130 cm^{-1} of transition v=3 → v=4 was found. It was shown that with a negative chirp of a femtosecond radiation pulse the population of high-lying vibrational levels of ν_1 modes significantly increases.

At a multiphoton excitation of selected vibrational modes in the molecule an implementation of photochemical bond-selective processes (or group of bonds) is possible. However, this possibility is limited due to effect of rapid intramolecular energy redistribution (*IVR*) from the excited mode to other vibrations of the molecule [1]. It is therefore important to study the process of *IVR* for different vibrational levels of the excited molecule. The aim of this work was to study vibrational dynamics in bis (trifluoromethyl) molecule $(CF_3)_2CCO$, resulting from multiphoton excitation of the ν_1 =2194 cm^{-1} mode of the C=C=O bond vibration by resonance IR femtosecond radiation.

In the experiments the IR pump-probe method with a spectral analysis of the probing radiation was used. For this purpose the new IR detector, namely high-sensitive and low-noise linear array on the basis of CdHgTe, developed with the participation of the authors, was employed for the first time. Measurements were carried out in the gas phase at low pressure, which made it possible to completely avoid the influence of intermolecular interaction. The duration of the pump and probe pulses was 120 fs with a spectral width of ~230 cm^{-1}. The spectra of induced transmission/absorption ΔOD depending on the delay

* Corresponding author: laptev@isan.troitsk.ru

time Δt between pump and probe pulses were measured in the range of wavelengths 2240–2040 cm^{-1}.

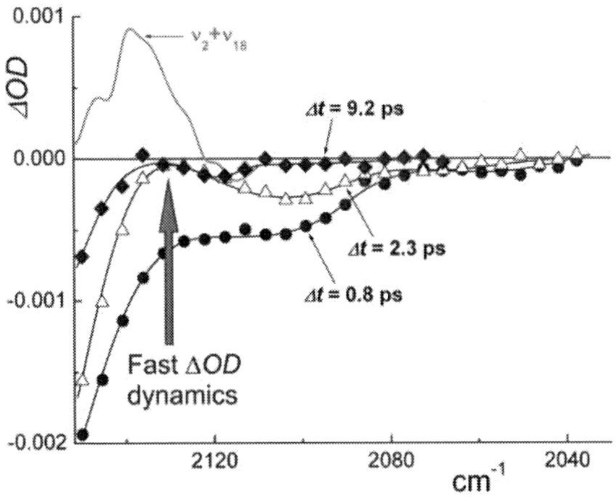

Fig. 1. "Tails" of *ΔOD* spectra in dependence on time delay Δt and fast dynamics at 2138 –2129 cm^{-1}.

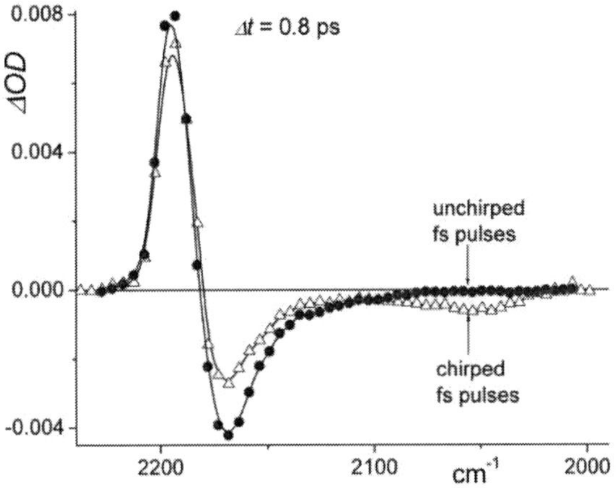

Fig. 2. *ΔOD* spectra for nonchirped (1.5×10^{12} W/cm^2) and negative chirped (1.8×10^{11} W/cm^2) laser femtosecond pulses.

The experiments confirmed a selective excitation of the resonant mode ν_1 up to vibrational level v = 5, which was previously registered by another way [2]. As the level of vibrational excitation of the resonant mode grows up the exponential decrease of the *ΔOD* signal was observed with characteristic times from τ = 6 ps for the transitions 0–1 and 1–2 to τ = 3.5 ps for the transitions 4–5 and 5–6. This corresponds to a reduction of *IVR* time from 6 to 1.2 ps in proportion to the quantum number v, which confirms the predictions of the theory [1]. At the same time, a sharp decrease of time τ to 1 ps at wavelengths of 2138–2129 cm^{-1}, approximately corresponding to the vibrational transition v=3→v=4 was found (see Fig. 1), which may be due to the resonance of the frequency of the excited vibrational level with the frequency of composite vibration $\nu_2+\nu_{18}$. Increasing the fluence/intensity of femtosecond pulses causes close to a proportional the growth of signal *ΔOD* and has virtually no effect on the dynamics of long "tail" of *ΔOD* spectrum. With a negative chirp of femtosecond radiation, when the frequency of the pulse decreases with time and thus adjusts to reduction of frequency of successive vibrational transitions due to the anharmonicity [3], it was found that the population of the high-lying vibrational states of molecules has increased considerably, although the radiation intensity has decreased by ~ 8 times (see Fig. 2).

References

1. V.S. Letokhov (ed.), *Laser Spectroscopy of Highly Vibrationally Excited States* (Adam Hilger, Bristol, 1989)
2. S.V. Chekalin, V.O. Kompanets, V.B. Laptev, S.V. Pigul'sky, A.A. Makarov, E.A. Ryabov, Chem. Phys. Lett. **512**, 178 (2011)
3. T. Witte, T. Homung, L. Windhorn, D. Proch, R. de Vivie-Riedle, M. Motzkus, K.L. Kompa, J. Chem. Phys. **118**, 2021 (2003)

Exciton-thermal analysis of nonisothermal nucleation in solid solution of CuCl in glass

V.I. Leiman*, A.L. Ashkalunin, O.Yu. Derkacheva, and V.M. Maksimov

St. Petersburg State University of industrial technologies and design, 198095 St. Petersburg, Russia

Abstract. The work presents results of investigation of nucleation of CuCl nanocrystals in glass by method of exiton-thermal analysis (ETA) which is based on measuring the optical absorbance of CuCl phase in sample under heating up to the melting temperature of CuCl nanocrystals. The melting temperature of nanoparticle depends on its radius. Therefore analysis of curve of melting CuCl phase allows to determine distribution curve of nanoparticles on their radii. By the method of ETA the features of variation of CuCl nanoparticles distributions in glass when changing regime of sample annealing (nonisothermal nucleation) were studied. For example, by reducing the heating rate of the sample the concentration of particles of CuCl phase increases considerably while reducing average radius.

Method of exiton-thermal analysis (ETA) of melting-crystallization of CuCl nanoparticles in glass is based on application of the well known Urbach's rule [1]. The rule reads: the long-wavelength absorption edge of crystals at different temperatures intersects in one "node point". The Urbach's rule is true for most of transparent media [2].

If you heat the sample and transmit light with energy of "node point" photon through the sample of glass with CuCl nanocrystals (NC), then absorbance NC CuCl does not change until beginning NC melting. Melting CuCl NC leads to transition of the nanoparticles in the amorphous state and the exciton states, which ensure absorption in the "nodal point", disappear. But zone-zone absorption of CuCl nanomelt remains. This absorption is almost two times less than nanocrystals one. The temperature of melting CuCl NC decreases with decreasing particles radius. Since NC distribution of radius takes a place in the glass, first the smallest CuCl particles of the distribution are melted, and then with increasing temperature the nanocrystals of bigger radii are melted. Thus, there is an optical scanning melting distribution of CuCl NC [3].

As a result, we have experimental curve of melting CuCl nanoparticles – dependence of optical absorption coefficient of the sample on temperature $K_0(T)$. Derivation $K_0(T)/dT$ is intensity of melting CuCl phase in dependence on sample temperature at linearly increasing temperature. Since melting NC of determined radius takes place at corresponding temperature, it is possible to obtain from the function $K_0(T)/dT$ the distribution of CuCl phase in sample on radii - the curve $f_V(r)=dV/dR$.

* Corresponding author: leiman@inbox.ru

Because ETA method is based on optical absorption it does this method very sensitive to investigate the CuCl nanocrystals in glass. ETA method allows to carry out studying CuCl NC with sizes from 1 to 30 nm and resolution up to 0.1 nm.

This method was applied to investigate features of process of nonisothermal nucleation of CuCl phase in glass at different rate of heating glass sample. It has been shown that a significant increasing the concentration of the CuCl phase is observed at decelerated heating (curves 1-6) sample compared with the rapid heating (curves 7-9). During slow heating the sample there is a maximum concentration of the CuCl phase in the 40th minute (curve 2). With the continuation of the nucleation (solution depletion) and the growth of temperature the critical radius increases, so that dissolving portion of previously formed CuCl clusters begins [4]. Therefore, there is a reduction CuCl phase concentration (curves 3-6).

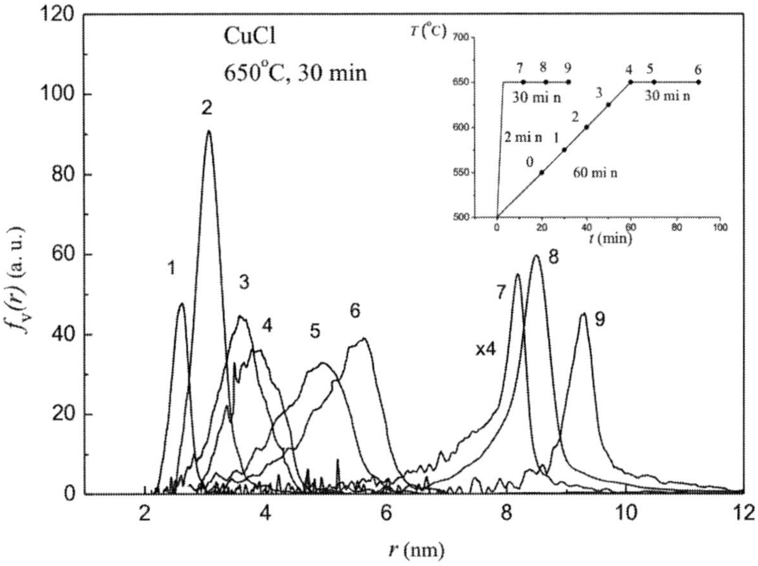

Fig. 1. Change in the distribution of the CuCl phase volume on the radius of the particle at slow heating (curves 1-6) and rapid heating (curves 7-9) the samples up to a temperature of 650 °C and holding for 30 minutes. Annealing time from the start of sample heating is shown in the inset by points and corresponds to the distribution number. The values of the curves 7-9 are multiplied by 4 times.

Thus, the first it is shown that at changing rate of heating a solid solution it should take into account the initial processes of nonisothermal nucleation which lead to a change in the sizes and concentration of new phase particles.

References

1. F. Urbach, Phys. Rev. **92**, 1324 (1953)
2. M.V. Kurik, Phys. Stat. Sol. A **8**, 9 (1971)
3. P.M. Valov, V.I. Leiman, Physics of the Solid State **51**, 1703 (2009)
4. P.M. Valov, V.I. Leiman, V.M. Maksimov, O.Yu. Derkacheva, Physics of the Solid State **52**, 821 (2010)

Electronic structure and luminescence properties of $Ca_2Ge_7O_{16}:Dy^{3+}$

I.I. Leonidov[1,2,*], A.V. Ishchenko[2], E.I. Konstantinova[1], V.P. Petrov[2], V.A. Chernyshev[2], and A.E. Nikiforov[2]

[1]Institute of Solid State Chemistry UB RAS, 620990 Ekaterinburg, Russia
[2]Ural Federal University, 620002 Ekaterinburg, Russia

Abstract. The present report represents an overview of the results of a combined experimental-computational study of electronic structure, thermoluminescence (TL) and afterglow properties of $Ca_2Ge_7O_{16}:Dy^{3+}$ synthesized for the first time. Afterglow curves of $Ca_2Ge_7O_{16}:Dy^{3+}$ at 575 nm showing persistent luminescence have been described in Becquerel law. The TL measurements reveal at least one TL band at 326 K and two luminescence bands at 475 and 535 nm. Persistent luminescence in $Ca_2Ge_7O_{16}:Dy^{3+}$ originates from relatively shallow charge traps with high probability of charge carriers recapture. The model of energy processes, configurations of traps and luminescence centers has been proposed with the aid of *ab initio* calculations performed using the LCAO approximation and several hybrid functionals.

Calcium and cadmium heptagermanates ($M_2Ge_7O_{16}$, M = Ca, Cd) are considered for development of multifunctional materials. The compounds have been extensively studied for their applications in anodes for lithium-ion batteries with excellent cycling performance, high capacity and superior rate capability [1]. Meanwhile, lanthanide-doped $M_2Ge_7O_{16}$ (M = Ca, Cd) demonstrate the possibility of their use as promising optical storage devices [2, 3].

The optical band gap of 5.50 eV in the $Ca_2Ge_7O_{16}$ (space group $P\text{–}4b2$, $Z = 2$) host has been evaluated from the diffuse reflection spectrum employing the Kubelka-Munk approximation. The band structure, density of states and partial density of states have been obtained from the results of DFT calculations performed using the LCAO approximation and several hybrid functionals which have shown the closest E_g^{opt} value (5.97 eV) to the experimentally estimated counterpart when the WC1LYP functional [4] has been applied.

While the highest intensity of photoluminescence in $Ca_{2-3x}Dy_{2x}Ge_7O_{16}$ in the visible spectral range is attained at $x = 0.0075$, $Ca_{1.985}Dy_{0.010}Ge_7O_{16}$ reveals the most pronounced orange afterglow in intensity and decay time. Persistent luminescence in $Ca_{1.985}Dy_{0.010}Ge_7O_{16}$ is observed for 60 min under 230 nm UV excitation ($E^{ex} = 5.39$ eV) for 5 min. Longer excitation time stimulates the detectable afterglow in $Ca_{1.985}Dy_{0.010}Ge_7O_{16}$ up to few hours.

* Corresponding author: ivanleonidov@ihim.uran.ru

The energy of UV excitation of $Ca_{2-3x}Dy_{2x}Ge_7O_{16}$ (5.39 eV) is approximately equal to the value of the optical band gap (5.50 eV). The presence of defects in the structure of $Ca_{2-3x}Dy_{2x}Ge_7O_{16}$ ($0 \leq x \leq 0.0075$) leads to additional acceptor levels in the forbidden gap near the bottom of the conduction band. The incident UV irradiation with the energy of 5.39 eV can be assumed to excite directly the corresponding trap levels. Duration of the observed afterglow depends on the energy of traps in the band gap. The maximal energies of the relatively shallow traps (1.00 eV and 0.72 eV) have been derived from the standard fitting of the measured thermoluminescence glow curves [5].

A possible mechanism of persistent luminescence with CIE chromaticity coordinates $x = 0.529$, $y = 0.429$ has been proposed for $Ca_2Ge_7O_{16}:Dy^{3+}$ that is similar to energy migration scheme discussed in [3]. Under UV irradiation, the excited electrons from the ground level of the host are captured by the native traps generated in $Ca_2Ge_7O_{16}$. During thermal disturbance at room temperature, captured electrons can be easily released to create the blue afterglow from the host. In dysprosium-doped $Ca_2Ge_7O_{16}$, the energy from the host emission transfers to excite energy levels of Dy^{3+}. The traps introduced by doping with Dy^{3+} form a bridge between the host and Dy^{3+} that promotes persistent energy transfer from the host to Dy^{3+} after the removal of exposure source. The shallow traps can also serve as trapping centers which distribute around the $4f$ states of Dy^{3+} near the bottom of the conduction band, therefore, more captured carriers transfer to Dy^{3+} after the removal of exposure source.

This work was partially supported by the FASO Program No. 0397–2015–0020 and RFBR Grant No. 14–03–31324. Computational procedures were carried out at the URAN Computing Platform at IMM UB RAS (Ekaterinburg, Russia). I.I.L. would like to acknowledge support from the Russian President Fellowship SP–931.2016.1.

References

1. Z.L. Hu, S. Zhang, C.J. Zhang, G.L. Cui, Coord. Chem. Rev. **326**, 34 (2016)
2. S.J. Yi, Y.L. Liu, J.X. Zhang, D.S. Yuan, J.H. Rong, L.H. Huang, Chinese J. Inorg. Chem. **20**, 247 (2004)
3. T. Wang, J. Gou, X.H. Xu, D. Zhou, J.B. Qiu, X. Yu, Opt. Express **23**, 12595 (2015)
4. R. Demichelis, B. Civalleri, M. Ferrabone, R. Dovesi, Int. J. Quant. Chem. **110**, 406 (2010)
5. R. Chen, S.W.S. McKeever, *Theory of thermoluminescence and related phenomena* (World Scientific Pub. Co., Singapore, 1997)

 DOI: 10.1051/epjconf/201713203028

Dielectronic recombination rate in statistical model

A.V. Demura, D.S. Leontyev*, V.S. Lisitsa, and V.A. Shurigyn

National Research Center "Kurchatov institute", 123182 Moscow, Russia

Abstract. The dielectronic recombination rate of multielectron ions was calculated by means of the statistical approach. It is based on an idea of collective excitations of atomic electrons with the local plasma frequencies. These frequencies are expressed via the Thomas-Fermi model electron density distribution. The statistical approach provides fast computation of DR rates that are compared with the modern quantum mechanical calculations. The results are important for current studies of thermonuclear plasmas with the tungsten impurities.

The dielectronic recombination (DR) plays the important role in the establishing ionization equilibrium of multielectron ions in high temperature plasma. In the modern thermonuclear facilities, tungsten is used for the elements of the discharge camera. Thus, the DR rate calculations of its ions are significant for evaluation of its impurities. The detailed level-by-level calculations of the DR rate dependence $Q_{dr}(T)$ on plasma temperature T for multielectron ions are a very difficult problem since for complex atoms like tungsten they require to take into account a huge number of atomic states and configurations [1]. The general expression for the DR rate represents itself the sum over branching ratios assembled from the total radiation $W_R(i)$ and autoionization $W_a(\mathrm{i},nl)$ rates of initial doubly excited states "i" and the partial autoionization rates $W_a(\mathrm{i},nl;f)$ [1].

$$Q^{P}{}_{DR}(T) = a_0^3\left(\frac{4\pi Ry}{T}\right)^{3/2}\sum_{inl,f}\frac{g_i}{g_f}\exp\left(-\frac{\hbar\omega_{if}}{T}+\frac{Z_i^2 Ry}{n^2 T}\right)\cdot\frac{W_a(\mathrm{i},nl;f)\cdot W_R(i)}{W_R(i)+W_a(\mathrm{i},nl)}, \tag{1}$$
$$W_R(i)\equiv\sum_{f'}W_R(i,\mathrm{nl};f',nl),\quad W_a(\mathrm{i},nl)\equiv\sum_{f''}W_a(\mathrm{i},nl;f''),$$

where a_0 is the Bohr radius, $2Ry$ is the atomic energy unit, Z_i is charge of the ion, g_i, g_f are the statistical weights of the initial and final states of ion core. The autoionization states are described by the sets of quantum numbers of the initial states of the ion core (i) and the ones of external electrons (nl). The set of the quantum numbers (f) describes the final stationary states of the ion core.

In the statistical approach of the multielectron ion is represented by a system of harmonic oscillators with the local plasma frequencies, expressed via the Thomas-Fermi electron density distribution $n_{TF}(\mathrm{r};Z_i,Z)$ (r is radius vector from nucleus) [2-3]

* Corresponding author: leontievdmitiy@gmail.com

$$\omega_p = \left(4\pi e^2 n_{TF}(\mathrm{r}; Z_i, Z)/m\right)^{1/2}, \tag{2}$$

where Z_i is the ion charge and Z is the charge of ion nucleus, e,m are the charge and mass of electron respectively. This model allows to describe universally different chemical elements. The partial radiation decay rates $W_R(i,\mathrm{f})$ could be expressed via the oscillator's strengths f_{if} (c is the speed of light) [4]

$$W_R(i,\mathrm{f}) = 2 f_{if}\omega_{if}/c^3. \tag{3}$$

The partial autoionization decay rate $W_a(\mathrm{i},nl;f)$ could be represented via the averaged over resonances near threshold the electron excitation cross-section $\sigma_{\mathrm{exc}}(\mathrm{f};\mathrm{i},\mathrm{n}\,l)$ of the doubly excited state (i,nl) [1, 4-5]

$$(2l+1)g_f W_a(\mathrm{i},nl;f) = Z_i^2 n^{-3}\omega_{if} g_i \sigma_{\mathrm{exc}}(\mathrm{f};\mathrm{i},\mathrm{n}\,l)/\pi^2. \tag{4}$$

For the simplification of the calculations it is conventional to average $W_a(\mathrm{i},nl;f)$ over the orbital momentum l of the external electron [1, 4-5], that gives with the use of quasiclassical asymptotics the result $W_a(\mathrm{i},n;\mathrm{f})$ [5]

$$W_a(\mathrm{i},n;\mathrm{f}) = < W_a(\mathrm{i},nl;f) >_l \approx (4\cdot 1.7/3\pi)\cdot(Z_i^2 f_{fi}/n^5\omega_{if}). \tag{5}$$

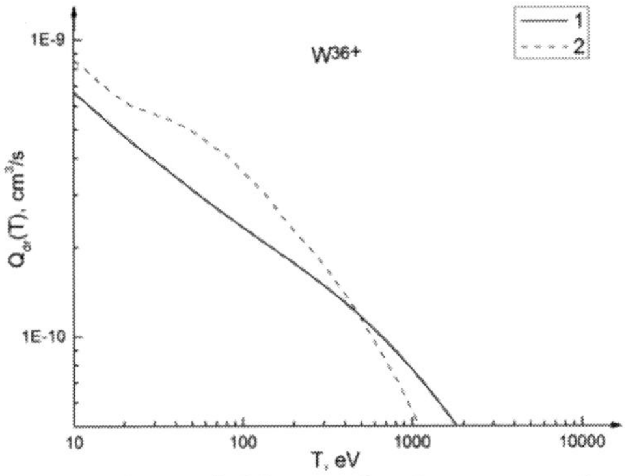

Fig. 1. The total DR rate for the tungsten ion W^{36+} as the function of electron temperature. Curve 1 is the calculation by statistical model; curve 2 is the data of detailed calculation [6].

As in the statistical approach, the oscillator's strengths also are expressed as the functions of electron density distribution the total DR rate is represented at the end as the functional of electron density.

The comparison of the total DR rates for the tungsten ion W^{36+}dependences versus T, calculated within the statistical approach and by the level-to-level package of codes [6], where finally the DR was evaluated using code FAC [7, 8], is presented in Figure 1. As one can see, there is a good agreement between the statistical model and detailed calculations.

As it could be seen from presented comparison the statistical approach provides the universal description of the DR rates for multielectron ions and allows to facilitate considerably its numerical evaluation.

References

1. I.I. Sobel'man, L.A. Vainstein, E.A. Yukov, *Excitations of Atoms and Broadening of Spectral Lines* (Springer, Berlin, 1995)
2. A.V. Demura, M.B. Kadomtsev, V.S. Lisitsa, V.A. Shurygin, Atoms **3**, 162 (2015)
3. W. Brandt, S. Lundqvist, Phys. Rev. **139**, 612 (1965)
4. W.H. Tucker, R.J. Gould, Astrophys. J. **144**, 244 (1966)
5. L.A. Bureyeva, T. Kato, V.S. Lisitsa, Phys. Rev. A **65**, 032702 (2002)
6. Z. Wu, C. Dong, *Theoretical Study on Electron Impact Excitation and Dielectronic Recombination of Highly Charged Tungsten Ions* (IAEA, Vienna, 2014)
7. M.F. Gu, Astrophys. J. **590**, 1131 (2003)
8. M.F. Gu, Can. J. Phys. **86**, 675 (2008)

Cooperative effects and slow dynamics of fluorescence intensity from quantum emitters in a dielectric

N.A. Lozing[1,2,*], and M.G. Gladush[1,2,3]

[1]Institute for Spectroscopy RAS, 108840 Moscow, Troitsk, Russia
[2]Moscow Institute of Physics and Technology, 141700 Dolgoprudny, Russia
[3]Moscow State Pedagogical University, 119991 Moscow, Russia

Abstract. We study theoretically the possibility of spontaneous switching between dim and bright fluorescence modes from a cooperative ensemble of two-level atoms driven by a cw-laser. A numerical analysis of transient regimes and transformations of the fluorescence spectrum are reported.

We study a bistability of two-level light emitters that are collected in a cooperative ensemble. It is known from the theories and experiments that many-particle systems may exhibit collective behavior because of interactions, like dipole-dipole, that are allowed or enhanced under certain conditions. Emission of light from such ensembles may be different than that from an uncorrelated group. Coherent spontaneous emission, superradiance, collective fluorescence and accompanying effects have been studied for several decades. At the same time a great number of problems specific for many-particle systems remain a challenge for theorists. The problem we apply to in this work is the fact that any cooperative system is likely to become optically bistable [1, 2]. In other words, there may be conditions for multiple steady-state populations and polarizations of the particles, realizable for the same external parameters. The presence of the material that hosts the light emitters brings changes and must be considered carefully even if it is a pure dielectric and any other interactions, like phonons, are neglected using valid approximations [3].

We use a generalized system of optical Bloch equations derived for an emitter in an ensemble of motionless similar particles embedded in a dielectric medium, which is transparent for the incident light [3-4]. The system is written with the effective rates of individual and collective radiative relaxation rates of the emitters, the Rabi frequency, and the effective transition-frequency shift caused by the presence of the dielectric host. The system's parameters are functions of the real and imaginary parts of the host's permittivity. The steady state solutions give hysteresis loops for the excited-state population of a two-level emitter if written as functions of the cw-laser field strength as shown in Fig. 1a. This population determines the total intensity of fluorescence and demonstrates the possibilities for switching between dim and bright fluorescence modes. If tuned near the hysteresis threshold points the system may be perturbed by either a fluctuation of the laser field or a small change in the density of the emitters. This perturbation is then likely to trigger the

* Corresponding author: lozing@phystech.edu

switching. The transient regimes were simulated by solving the equations numerically. It appeared to be a long-time switching process and it was shown to be orders of magnitude (10^4 and may be longer) longer than the excitation lifetime (see Fig. 1. b and c). When switching between the dim and the bright fluorescence occurs some transformations of the fluorescence spectrum are obviously expected. It was shown to be so by calculations of the fluorescence spectra both analytically and numerically. We assume, that the discussed effects could be observed experimentally by photon echo technique [5-7] or various methods of laser site-selective spectroscopy at low temperatures [8].

It appeared to be a long-time switching process and it was shown to be orders of magnitude (10^4 and may be longer) longer than the excitation lifetime (Fig. 1. b and c).

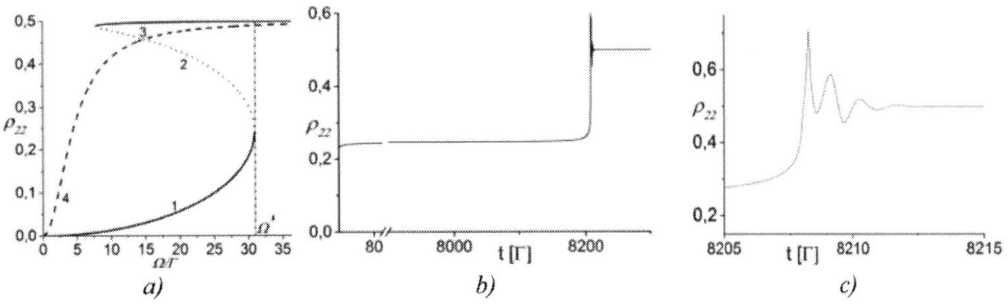

Fig. 1. a) Steady state population of excited state, b) time-dependence of population after perturbing, c) time-dependence of population after perturbing (increased scale) near the transition point.

This work was supported by RFBR (proj. no. 16-02-01174).

References

1. L. Mandel, E. Wolf, *Optical Coherence and Quantum Optics* (Cambridge University Press, New York, 1995)
2. M.G. Gladush, D.V. Kuznetsov, V.K. Roerich, Eur. Phys. J. D **64**, 511 (2011)
3. D.V. Kuznetsov, V.K. Roerich, M.G. Gladush, J. Exp. Theor. Phys. **113**, 647 (2011)
4. M.G. Gladush, EPJ Web of Conferences **103**, 02004 (2015)
5. Y.G. Vainer, A.V. Naumov, M.A. Kol'chenko, R.I. Personov, Phys. Stat. Sol. B **241**, 3480 (2004)
6. K.R. Karimullin, M.V. Knyazev, A.V. Naumov, Bull. RAS: Physics **78**, 1254 (2014)
7. K.R. Karimullin et al., Meas. Sci. Technol. **24**, 027002 (2013)
8. A.V. Naumov, Physics–Uspekhi **56**, 605 (2013)

Luminescent microscopy of the small ensembles of the CdSe nanocrystalls synthesized in the liquid crystal matrix of the cadmium octanoate

K.A. Magaryan[1,*], I.Y. Eremchev[2], K.R. Karimullin[1,2], and I.A. Vasilieva[1]

[1]Moscow State Pedagogical University, 119435 Moscow, Russia
[2]Institute for Spectroscopy RAS, 108840 Troitsk, Moscow, Russia

Abstract. Here we present investigation of the thin film with CdSe nanocrystals with the size 2.3 nm made from low-concentration toluene solution. Using fluorescent microscopy method we have found small nanocrystals ensembles up to single nanocrystals. Luminescence spectra in different parts of the test sample were registered. Analysing them we can draw a conclusion about the inhomogeneous nature of the luminescence spectra of the bulk nanocomposite material.

Combination of the unique properties of the new functional nanostructures and the unique features of the liquid crystal (LC) materials expands possibilities for their potential use significantly. Today, there are three main approaches for obtaining composites based on LC materials with nanostructures. They are doping LC matrix with nanoparticles, spraying nanostructures on the LC substrate surface and in-situ method of the direct synthesis inside the LC material [1]. To investigate spectral characteristics of the 3rd type structures we were provided with the CdSe quantum dots (QDs) grown directly in the ionic thermotropic LC matrix [2].

Polycrystalline powder with CdSe QDs was dispersed in chemically pure toluene and deposited on a substrate. After that it was placed on a precise 2D scanning stage of the fluorescence microscope–spectrometer [3]. The XY travel range was 25 μm^2 with a step of 0.5 μm approximately. Luminescence spectrum was recorded on every step. All luminescence spectra were obtained at room temperature using DPSS laser diode pumping at wavelength of 405 nm.

Inside the red frame on the Fig. 1 there are several examples of the recorded spectra of the different diffraction-limited areas on the investigated region. As one can see these curves fall in the spectral region of a broad luminescence band (purple curve), which was previously associated with the decay of excitons localized on the intrinsic defects on the QD interface [4]. The peak at wavelength of 518 nm is the most frequently detected. It corresponds to the interband transition characteristic for nanocrystals with a size of 2.3 nm. The broad band in the region 575 – 750 nm is inhomogeneously broadened because it contains many wide and narrow individual lines. Some of these lines show the effect of blinking [5-9]. This blinking, as well as small FWHM (10 nm at the average) of these lines

* Corresponding author: ka.magaryan@phys.mpgu.edu

indicate that the light sources are QDs with larger diameter. Apparently QDs grew to various sizes at the stage of synthesis of nanocomposite material and now it gives contribution to the bulk fluorescence spectrum.

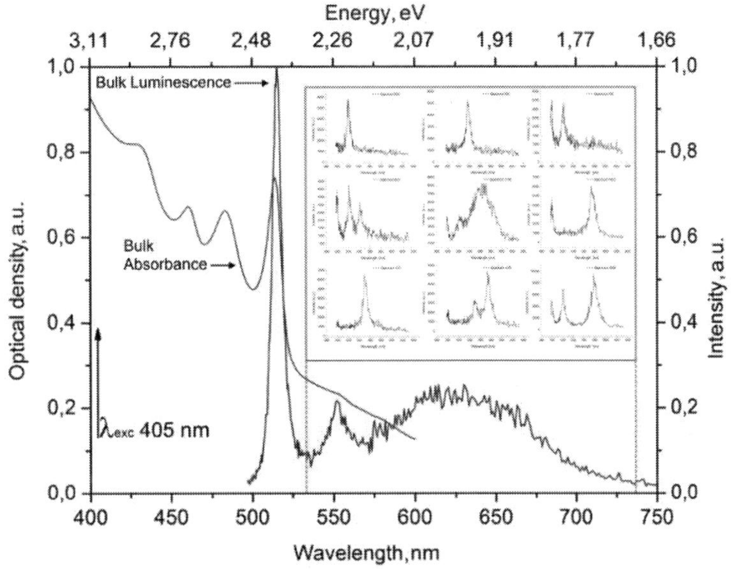

Fig. 1. Green solid curve - absorption spectrum of the toluene solution with 2.3 nm CdSe QDs. Purple solid curve is the fluorescence spectrum of the same solution registered from the bulk volume. Curves in the red frame show multiple luminescence spectra registered in the different parts of the thin film with CdSe QDs.

The authors are grateful to Prof. G.V. Klimusheva (Institute of Physics, National Academy of Sciences, Kyiv, Ukraine), Prof. T.A. Mirnaya (V.I. Vernadskii Institute of General and Inorganic Chemistry, National Academy of Sciences, Kyiv, Ukraine) for the opportunity to study sample with CdSe QDs, and Prof. Dr. A.V. Naumov for useful discussions.

The work is supported by the Russian Foundation for Basic Researches (project № 15-32-21100-mol_a_ved). Program of Presidium RAS "Fundamental and applied problems of Photonics and physics of new optical materials". K. Magaryan acknowledges the Ministry of Education and Science of the Russian Federation (project No 14.B25.31.0007).

References

1. Li Quan, *Nanoscience with Liquid Crystals* (Springer, USA, 2014)
2. D. Zhulai, D. Fedorenko, A. Kovalchuk, S. Bugaychuk, G.V. Klimusheva, T.A. Mirnaya, NRL **10,** 66 (2015)
3. A.V. Naumov, I.Yu. Eremchev, A.A. Gorshelev, Eur. Phys. J. D **68**, 348 (2014)
4. K.A. Magaryan, M.A. Mikhailov, K.R. Karimullin, M.V. Knyazev, I.Yu. Eremchev, A.V. Naumov, I.A. Vasilieva, G.V. Klimusheva, J. Lumin. **169**, 799 (2016)
5. T. Plakhotnik et al., Phys. Rev. Lett. **105**, 167402 (2010)
6. M.J. Fernee et al., J. Phys. Chem. Lett. **3**, 1716 (2012)
7. A.L. Shchukina, I.Yu. Eremchev, A.V. Naumov, Phys. Rev. E **92**, 032102 (2015)
8. I.S.Osad'ko, I.Yu. Eremchev, A.V. Naumov, J. Phys. Chem. C **119**, 22646 (2015)
9. I.Yu. Eremchev, I.S.Osad'ko, A.V. Naumov, J. Phys. Chem. C **120**, 22004 (2016)

DOI: 10.1051/epjconf/201713203031

Reemission spectra and inelastic processes at interaction of attosecond and shorter duration electromagnetic pulses with atoms

D.N. Makarov*, and V.I. Matveev

M.V. Lomonosov Northern (Arctic) Federal University, 163002 Arkhangelsk, Russia

Abstract. Inelastic processes and the reemission of attosecond and shorter electromagnetic pulses by atoms have been considered within the analytical solution of the Schrödinger equation in the sudden perturbation approximation. A method of calculations with the exact inclusion of spatial inhomogeneity of the field of an ultrashort pulse and the momenta of photons in the reemission processes has been developed. The probabilities of inelastic processes and spectra of reemission of ultrashort electromagnetic pulses by one- and many-electron atoms have been calculated. The results have been presented in the form of analytical formulas.

Progress currently achieved in the methods of generation of short and ultrashort electromagnetic pulses has allowed overcoming the "femtosecond limit" and obtaining pulses with a duration of several tens of attoseconds. As a result, attosecond physics appeared (see, e.g., reviews [1–5]) and it became possible to observe atomic phenomena on a real time scale. The processes accompanying the interaction of ultrashort pulses with atoms are usually described within the following approaches. For pulses shorter than the characteristic atomic time, but longer than attosecond pulses, it is appropriate to use approaches based on the Magnus expansion for the evolution operator. Only the first several terms in the expansion of the evolution operator [6] are currently used in calculations. In these calculations, the field of the pulse is assumed to be spatially uniform at the dimensions of a target [7, 8]. The sudden perturbation approximation is appropriate for attosecond and shorter pulses. The possibilities of generation of pulses much shorter than attosecond ones are actively discussed (see, e.g., [9, 10]). The exact solution of the Schrödinger equation in the sudden perturbation approximation for the interaction of attosecond and shorter electromagnetic pulses with multielectron atoms has been obtained in [11]. It is usually accepted [12] that the integral of the electric field strength of a laser source in vacuum with respect to time is zero. The description of the interaction of an atomic electron in the sudden perturbation approximation implies that the electron is free during the action of the field of the pulse; consequently, the absorption of laser radiation in vacuum is significantly nonlinear with respect to the field. The exact solution obtained in [11] makes it possible to calculate such effects.

* Corresponding author d.makarov@narfu.ru

In this work, a method for calculating inelastic processes and spectra of reemission of extremely short electromagnetic pulses by atoms is developed on the basis of the exact solution of the Schrödinger equation in the sudden perturbation approximation obtained in [11]. In the cases under consideration, it is assumed that the duration of such pulses is much shorter than the characteristic atomic time. Thus, we consider attosecond and shorter electromagnetic pulses. The developed method allows the exact inclusion of the spatial inhomogeneity of the field of the electromagnetic pulse and the momenta of photons in the processes of reemission. The dipole approximation is not used. The field of the incident pulse is taken into account exactly in the sudden perturbation approximation, whereas the emission of a photon is described in perturbation theory. It should be noted that we are speaking of the emission of one photon by all atomic electrons during the action of a sudden perturbation. The reemission spectra include incoherent (proportional to N) and coherent (proportional to N2) parts, where N is the number of electrons in the atom. The results are presented as simple analytical formulas containing several coefficients and screening parameters tabulated [13] for all atoms (with nuclear charges $1 \leq Z \leq 92$) with electron densities described by the well-known Dirac–Hartree–Fock–Slater model. By way of example, the spectra of reemission of ultrashort electromagnetic field Gaussian pulses by carbon, silicon, germanium, tin, and lead atoms are calculated. The method can be used not only for a Gaussian pulse, but also for any other pulse, provided that the sudden perturbation approximation is applicable. The choice of the Gaussian pulse is convenient because it allows following the transition from the field whose the integral of the electric field strength of the ultrashort pulse with respect to time is nonzero to the field with zero integral at the passage to limit. As an illustration, we numerically calculated the total probability of all inelastic processes in the hydrogen-like atom by using Gaussian pulse. Obviously, our method also permits direct generalization to molecular targets.

This work was supported by the Ministry of Education and Science of the Russian Federation (state assignment no. 3.1726.2014/K) and by the Council of the President of the Russian Federation for Support of Young Scientists and Leading Scientific Schools (project no. SP-1800.2015.1).

References

1. F. Krausz, M. Ivanov, Rev. Mod. Phys. **81**, 163 (2009)
2. P.P. Agostini, L.F. DiMauro, Rep. Prog. Phys. **67**, 813 (2004)
3. P.V. Corkit, F. Krausz, Nat. Phys. **3**, 381 (2007)
4. V. Astapenko, *Interaction of Ultrashort Electromagnetic Pulses with Matter, Springer Briefs in Physics* (Springer, Heidelberg, 2013)
5. A.M. Zheltikov, Phys. Usp. **54**, 29 (2011)
6. A.M. Dykhne, G.L. Yudin, Sov. Phys. Usp. **21**, 549 (1978)
7. M. Klaiber, D. Dimitrovski, J.S. Briggs, Phys. Rev. A **79**, 043402 (2009)
8. D. Dimitrovski, M. Ferre, L.B. Madsen, Phys. Rev. A **80**, 053412 (2009)
9. A.E. Kaplan, Lasers Eng. **24**, 3 (2013)
10. G. Mourou, S. Mironov, E. Khazanov, A. Sergeev, Eur. Phys. J. **223**, 1181 (2014)
11. D.N. Makarov, V.I. Matveev, JETP Lett. **103**, 415 (2016)
12. B.M. Karnakov, V.D. Mur, S.V. Popruzhenko, V.S. Popov, Phys. Usp. **58**, 3 (2015)
13. F. Salvat, J.D. Martinez, R. Mayol, J. Parellada, Phys. Rev. A **36**, 467 (1987)

EPJ Web of Conferences **132**, 03032 (2017) DOI: 10.1051/epjconf/201713203032
SPECTROSCOPY.SU 2016

Hyperfine structure of S-states of muonic tritium

F.A. Martynenko[1,*], R.N. Faustov[2], and A.P. Martynenko[1]

[1]Samara University, 443086 Samara, Russia
[2]Institute of Informatics in Education, FRC CSC RAS, 119333 Moscow, Russia

Abstract. On the basis of quasipotential method in quantum electrodynamics we carry out a precise calculation of hyperfine splitting of S-states in muonic tritium. The one-loop and two-loop vacuum polarization corrections, relativistic effects, nuclear structure corrections in first and second orders of perturbation theory are taken into account. The contributions to hyperfine structure are obtained in integral form and calculated analytically and numerically. Obtained results for hyperfine splitting can be used for a comparison with future experimental data of CREMA collaboration.

A precise calculation of hyperfine splitting of S-states in muonic tritium atom is carried out. Muonic tritium consists of a triton and a negatively charged muon. Its lifetime is determined by the lifetime of the muon τ=2.19703940·μs. Measuring the frequency of transition (2P-2S) in muonic hydrogen, helium ions with an accuracy of 50 ppm will yield the value of the charge radius of the proton, deuteron, triton, helion and α-particle with an accuracy of 0.0005 fm [1,2]. For comparison between theory and experiment it is necessary to perform analytical and numerical calculation of the corrections of the fifth and sixth order over the fine structure constant in the hyperfine structure (HFS) of S-states of muonic tritium within the quasipotential method in quantum electrodynamics. Muon atoms differ from electron atoms that for them there is a wide class of corrections to the energy spectrum, which should be examined directly for each muon atom.

On the basis of quasipotential method [3-5] the two-particle bound state is described by the Schrödinger equation. The main contribution to the interaction operator of a muon and triton is determined by the Breit Hamiltonian:

$$H_B = H_0 + \Delta V_B^{fs} + \Delta V_B^{hfs}, \quad H_0 = \frac{p^2}{2\mu} - \frac{Z\alpha}{r}, \tag{1}$$

$$\Delta V_B^{fs} = -\frac{p^4}{8m_1^3} - \frac{p^4}{8m_2^3} + \frac{\pi Z\alpha}{2}\left(\frac{1}{m_1^2} + \frac{1}{m_2^2}\right)\delta(\mathbf{r}) - \frac{Z\alpha}{2m_1m_2r}\left(p^2 + \frac{\mathbf{r}(\mathbf{r}\mathbf{p})\mathbf{p}}{r^2}\right), \tag{2}$$

$$\Delta V_B^{hfs} = \frac{8\pi\alpha\mu_t}{3m_1m_p}\frac{\boldsymbol{\sigma}_1\boldsymbol{\sigma}_2}{4}\delta(\mathbf{r}), \tag{3}$$

where m_1, m_2 - masses of muon and triton, m_p - proton mass, μ_t - triton magnetic moment.

The potential (3) gives the leading order contribution to hyperfine structure (the Fermi energy). One part of corrections to hyperfine structure is known in analytical form from calculations in electronic hydrogen [6]. They can be used to estimate the contributions to the HFS of muonic tritium. But there are effects that are specific to each of the muon atom.

* Corresponding author: f.a.martynenko@gmail.com

One such effect is the effect of vacuum polarization, which leads to an additional term in the Hamiltonian

$$\Delta V^{hfs}_{1\gamma,VP}(r) = \frac{8\alpha\mu_t}{3m_1 m_p}\frac{\boldsymbol{\sigma}_1\boldsymbol{\sigma}_2}{4}\frac{\alpha}{3\pi}\int_1^\infty \rho(s)ds\left[\pi\delta(r) - \frac{m_e^2\xi^2}{r}e^{-2m_e sr}\right], \tag{4}$$

The potential (4) gives the contribution to HFS of order α^5 after averaging over the wave functions. The same order contribution is determined by muon anomalous magnetic moment. The two-loop vacuum polarization potentials are also considered. In this work we take into account one-loop and two-loop corrections on the vacuum polarization in first and second orders of perturbation theory. Relativistic corrections of six order in α are included in total value of hyperfine splitting.

Nuclear structure and recoil effects are determined in the leading order by two-photon exchange amplitudes. For their calculation the triton vertex operator is parameterized by two electromagnetic form factors, and the contribution of these amplitudes to the hyperfine structure is represented in integral form and calculated numerically. To construct the interaction operator from two-photon exchange amplitudes with the account of radiative corrections we use a method of projection operators on two-particle states with definite spin S=0,1. The introduction of such projection operators allows further to use package Form [7] to calculate the trace of the product of the Dirac matrices and convolutions by Lorentz indices [8-9]. Numerical contributions of 2γ-amplitudes to HFS are obtained by means of the Gaussian and dipole parameterizations for the triton form factors. The most significant theoretical error is related namely with 2γ-amplitudes. We calculate also more complex corrections which include both nuclear structure and vacuum polarization effects.

To calculate different corrections in second order perturbation theory we use a compact representation for the reduced Coulomb Green's function obtained in [10]. To calculate numerous matrix elements of interaction operator (coordinate integration) we use a system Mathematica. The values of fundamental physical constants (particle masses, fine structure constant, muon anomalous magnetic moment, triton charge radius) are taken from [11-12]. Total more precise values of hyperfine splittings are obtained: (237.667±0.085) meV (1S-state) and (29.703±0.011) meV (2S-state), which can be used for a comparison with future experimental data of CREMA (Charge Radius Experiments with Muonic Atoms) collaboration [1-2]. Performed precise calculation of the hyperfine structure of the muonic tritium is also important to calculate the formation rate of mesic molecules (ptμ), (dtμ) and flow of muon catalyzed reactions [13].

The support from Russian Foundation for Basic Research (grant 16-02-00554) and Ministry of Education and Science of Russia (grant 1394) is acknowledged.

References

1. R. Pohl, A. Antognini, F. Nez et al., Nature **466**, 213 (2010)
2. J.J. Krauth, M. Diepold, B. Franke et al., Ann. Phys. **366**, 168 (2016)
3. A.P. Martynenko, J. Exp. Theor. Phys. **106**, 690 (2008)
4. R.N. Faustov, A.P. Martynenko, J. Exp. Theor. Phys. **88**, 672 (1999)
5. R.N. Faustov, A.P. Martynenko, J. Exp. Theor. Phys. **98**, 39 (2004)
6. M.I. Eides, H. Grotch, V.A. Shelyuto, Phys. Rep. **342**, 62 (2001)
7. J.A.M. Vermaseren, e-preprint arXiv:math-ph/0010025
8. R.N. Faustov et al., Phys. Lett. B **733**, 354 (2014)
9. R.N. Faustov et al., Phys. Rev. A **90**, 012520 (2014)
10. H.F. Hameka, J. Chem. Phys. **47**, 2728 (1967)
11. P.J. Mohr, B.N. Taylor, D.B. Newell, Rev. Mod. Phys. **84**, 1527 (2012)
12. Angeli, K.P. Marinova, Atom. Data Nucl. Data Tables **99**, 69 (2013)
13. S.S. Gershtein, Yu.V. Petrov, L.I. Ponomarev, Sov. Phys. Usp. **33**, 591 (1990)

Selection of the optimal combination of water vapor absorption lines for detection of temperature in combustion zones of mixing supersonic gas flows by diode laser absorption spectrometry

V.R. Mironenko[*], Yu.A. Kuritsyn, M.A. Bolshov, V.V. Liger

Institute for Spectroscopy RAS, 108840 Troitsk, Moscow, Russia

Abstract. Determination of a gas medium temperature by diode laser absorption spectrometry (DLAS) is based on the measurement of integral intensities of the absorption lines of a test molecule (generally water vapor molecule). In case of local thermodynamic equilibrium temperature is inferred from the ratio of the integral intensities of two lines with different low energy levels. For the total gas pressure above 1 atm the absorption lines are broadened and one cannot find isolated well resolved water vapor absorption lines within relatively narrow spectral interval of fast diode laser (DL) tuning range (about 3 cm^{-1}). For diagnostics of a gas object in the case of high temperature and pressure DLAS technique can be realized with two diode lasers working in different spectral regions with strong absorption lines. In such situation the criteria of the optimal line selection differs significantly from the case of narrow lines. These criteria are discussed in our work. The software for selection the optimal spectral regions using the HITRAN-2012 and HITEMP data bases is developed. The program selects spectral regions of DL tuning, minimizing the error of temperature determination $\delta T/T$, basing on the attainable experimental error of line intensity measurement δS. Two combinations of optimal spectral regions were selected – (1.392 & 1.343 μm) and (1.392 & 1.339 μm). Different algorithms of experimental data processing are discussed.

We shall consider the problem on errors in evaluation of temperature [1-3]. It is possible to show, that a systematic error in evaluation of temperature

$$\frac{\Delta T}{T} = \frac{T}{T_{eff}}\frac{\Delta R}{R} = \frac{T}{T_{eff}}\left(\frac{1}{S_1}+\frac{1}{S_2}\right)^{1/2}\Delta S, \quad (1)$$

where $T_{eff} = \frac{hc\Delta E''}{k}$, $R = \frac{S_2}{S_1}$.

The error ΔT is proportional to T^2/T_{eff} and depends on relative errors in line strengths S_1 and

[*] Corresponding author: miron@isan.troitsk.ru

S_2 of the transitions. In turn, these errors are determined by an inaccuracy of the parameters in data bases, experimental measuring errors and methodical errors of data processing.

Besides the above mentioned errors of ΔT, there are the errors caused by noise in system (statistical errors). By analogy with (1), the expression for statistical error of evaluation of temperature (in the conjecture of independence and equality of statistical errors $\delta S_1 = \delta S_2 = \delta S$) can be written in the form:

$$\delta T = \frac{T^2}{T_{eff}}\left(\frac{1}{S_1^2}+\frac{1}{S_2^2}\right)^{1/2}\delta S = \frac{T^2}{T_{eff}}\left(\frac{1}{S_1^2}+\frac{1}{S_2^2}\right)^{1/2}\frac{\delta A}{NL}. \quad (2)$$

In (2) we used, that statistical error in measurements of line strength S of a transition is connected with an error in measurement of the integral absorption A of line in the form: $\delta S = \delta A/NL$. In a probing volume with total pressure P and relative density of molecules χ, $N = \chi P/kT$ and

$$\delta T = \frac{kT^3}{T_{eff}\chi PL}\left(\frac{1}{S_1^2}+\frac{1}{S_2^2}\right)^{1/2}\delta A = \frac{kT^3}{T_{eff}\chi PL}(1+R^2)^{1/2}\frac{\delta A}{S_1}. \quad (3)$$

Under the designed program two optimum combinations of spectral regions have been selected:

Combination 1: "cold" $\lambda = 1.392$ μm ($\nu = 7184 - 7187$ cm^{-1}), "hot" $\lambda = 1.343$ μm ($\nu = 7443 - 7446$ cm^{-1}):

Combination 2: "cold" $\lambda = 1.392$ μm ($\nu = 7184 - 7187$ cm^{-1}), "hot" $\lambda = 1.339$ μm ($\nu = 7465 - 7468$ cm^{-1}).

Calculated dependences of S and δT for two combinations over the temperature range 500 – 2000 K, expected in experiment are presented in the Figure. It is evident, that both combinations can be used for measuring temperature in a whole temperature range 500 – 2000 K, however the combination *1* ensures more precise evaluation of temperature over the range 500 – 1600 K, and the combination *2* – over the range 1600 – 2000 K, because the lines in spectral region $\nu = 7465 - 7468$ cm^{-1} start from "hotter" lower levels.

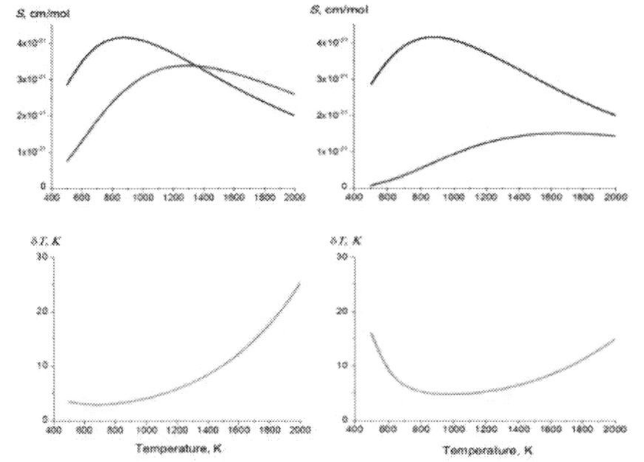

Fig. 1. Line strengths (the upper graphs) and errors of evaluation of temperature (the lower graphs) for two combinations of lines: the combination 1 (the left graphs), the combination 2 (the right graphs). Blue color is used for designation of "cold" lines, red color – for "hot" lines.

The support from the Program of the Department for General Physics of RAS "Basic problems of laser technologies" is acknowledged.

References

1. X. Zhou, X. Liu, J. Jeffries, R.K. Hanson. Meas. Sci. Technol. **14**, 1459 (2003)
2. M.A. Bolshov et al., Opt. Spectrosc. **110**, 848 (2011)
3. X. An et al., J. Quant. Spectrosc. Radiat. Transfer **112**, 2355 (2011)

Coherent spectroscopic methods for monitoring pathogens, genetically modified products and nanostructured materials in colloidal solution

T. Moguilnaya[1,2,*], Y. Suminov[1] , A. Botikov[3], S. Ignatov[4], A. Kononenko[5], and A. Agibalov[2]

[1]JSC Giprogazoochistka, 105203 Moscow, Russia
[2]JSC "SERVET-M", 115404 Moscow, Russia
[3]Research Institute of Epidemiology and Microbiology, 123098 Moscow, Russia
[4]State Research Center for Applied Microbiology and Biotechnology, 142279 Moscow, Russia
[5]Veterinary Sanitation, Hygiene and Ecology Institute, 119991 Moscow Russia

Abstract. We developed the new automatic method that combines the method of forced luminescence and stimulated Brillouin scattering. This method is used for monitoring pathogens, genetically modified products and nanostructured materials in colloidal solution. We carried out the statistical spectral analysis of pathogens, genetically modified soy and nano-particles of silver in water from different regions in order to determine the statistical errors of the method. We studied spectral characteristics of these objects in water to perform the initial identification with 95% probability. These results were used for creation of the model of the device for monitor of pathogenic organisms and working model of the device to determine the genetically modified soy in meat.

We developed the coherent spectroscopic method for the monitoring of the pathogenic organisms directly in water pipe-line, genetically modified products and nanostructured materials in colloidal solution. The method is based on an analysis of spectral characteristics of stimulated radiation, passed through the solution.

We studied transmission IR spectra of a number of pathogens (salmonella, viruses of herpes genitaler, hepatitis A and C, grippe A and B ecd) and nanosilver solutions and luminescence of nanomarkers. In our experiments, the laser radiation passed through a quartz cell with water solutions of the pathogens, nanosilver or nanomarkers. Spectra were analyzed by the spectrum analyzer "Agilent" (USA) with spectral resolution 0.5 nm. The solid DPSS lasers ("Laser-export Co. Ltd.", Russia) with wavelengths 1017, 810, 670, 532, 480 nm were used as the sources of exciting radiation. The input/output waveguides bring the exciting and passing radiations from laser to the cell and from cell to the analyzer, respectively.

We found that exciting radiations with wavelengths 1017 and 810 nm induce the stimulated Brillouin scattering in spectra of the water containing pathogen DNAs. We revealed that peak positions and widths of "fingerprints" for pathogens under study, and optical densities of these bands were proportional pathogen content, if their content was

* Corresponding author: mogilnay@mail.ru

less 15%. Thus, Stokes and anti-Stockes bands of the stimulated Brillouin scattering can be used to recognize the pathogens.

Nonlinear effects like SBS arising doe to the density of radiation in this solution [1]. The similar density of radiation is reached due to combination of the luminescence fields and laser radiation [2] giving the characteristic line. The modal frequency of this line depends on type DNA and accordingly from type of a pathogenic organism [3]. Similar lines are well interpreted by methods of the theory of recognition of images.

We have also established the new method for rapid analysis of genetically modified soy in meat products detected by a modified PCR fluorescent method. The transmission spectra of the pathogens, fluorescent markers, mix spectra (meat products, a soya) in a stationary solution and in the course of reaction have been investigated. It was found out that luminescence (around 1000 – 1300 nm) was defined only PCR reaction, thus, parasitic radiation from spontaneous luminescence of albumin was minimum. Luminescence at 1500 nm is defined only for albumin of meat. It is shown that besides a signal of the fluorescent marker in the course of carrying out PCR of reaction, the own fluorescence of the soya exists at 1000 nm. We have developed a working model of the device for the automated monitoring genetically modified sample soy. Based on these studies, we have developed the device for the automated control of pathogens in the water directly. We have also established the new method for rapid analysis of genetically modified products, representing a modified PCR fluorescent method.

The inventive methods can be recommended for DNA diagnosis, including field conditions, in medicine, veterinary sciences, in sanitary and epidemiological studies for detecting agents of dangerous infections, including potential bio-terrorist attacks.

It is of interest to develop this method to study changes in the composition under the influence of various factors. It was of interest to determine not only the content of the active cells, but also the content of destroyed cells. We are currently investigated spectral dynamics of invasive and inactivated pathogen. We measured the spectra of luminescence of live pathogens for several seconds, then the intensity of the luminescence decayed and we analyzed the peaks of the laser mode and the Stokes or anti-Stokes peaks. Stokes components were observed within 12 – 25 min depending on the type of pathogen, then only the peak of the laser mode was observed. We did not detect luminescence inactivated pathogen. In case the dead pathogens, after a delay of 2 – 3 min, the peak intensity of the Stokes component increased and the intensity of the laser mode decreased. We found that the intensity of the Stokes component for the dead bacterial cells was inactivated by several orders of magnitude to compare with alive bacterial cells toghether with reducing of the intensity of the laser mode.

We studied three pathogens – *Staphylococcus aureus*, *Escherichia coli* and coliphage MS-2. We found the Lamb dip in the peak corresponding to the Stokes and anti-Stokes component in all spectra of bacteriophages and did not reveal it in the virus spectrum.

We studied the influence of nanosilver on changes on the concentration of the live pathogens. After increase of the silver content, the peak was observed in the anti-Stokes field and the intensity decreased. We assume that this effect is associated with the death of pathogens. Standard biological method is qualitatively in agreement with our results.

We are currently investigated the spectral dynamics of particles of nanosilver depending on concentration. Spectral characteristics are changed at the nanoparticle concentration of 10 mg/l. It is likely due to the aggregation of the particles.

References

1. T.Yu. Moguilnaya, E.A. Saguitova, A.G. Botikov, Proc. of SPIE **5621**, 18 (2003)
2. T. Moguilnaya, P. Bobkov, V. Tomilin, J. Phys.: Conf. Ser. **414**, 012023 (2013)
3. T. Moguilnaya et al., Pribory **1**, 45 (2013) (in Russian)
4. T.Yu. Moguilnaya et al., J. Phys.: Conf. Ser. **594**, 012026 (2015)

EPJ Web of Conferences **132**, 03035 (2017) DOI: 10.1051/epjconf/201713203035
SPECTROSCOPY.SU 2016

The investigations of nanoclusters and micron-sized periodic structures created at the surface of the crystal and amorphous silica by resonant CO_2 laser irradiation

A.F. Mukhamedgalieva[1,*], A.M. Bondar[1], I.M. Svedov[1], M.A. Kononov[2], V.B. Laptev[3], and N.N. Novikova[3]

[1]National University of Science and Technology MISIS, 119049 Moscow, Russia
[2]Institute of General Physics RAS, 119991 Moscow, Russia
[3]Institute for Spectroscopy RAS, 108840 Troitsk, Moscow, Russia

Abstract. The creation of nanoclasters and micrometer sized periodical structures at the surface of silica (crystal quartz and fused quartz) by action of pulsed CO_2 laser radiation (pulse energy of 1 J, pulse time of 70 ns) have been investigated. The laser action on the surface of samples lead to appearance of two kind of structures – periodical micron-sized structures with the period length close to wave length of CO_2 laser irradiation and nanoclusters with size close to 50-100 nanometers. This creation connects with the intensive ablation of matter at the maxima of standing waves which are a results of the interference of falling and surfaces waves. This connects with the resonant absorption of infrared laser radiation by silicate minerals.

The appearance of nanoclasters and micrometer sized periodical structures at the surface of silica (crystal quartz and fused quartz) by action of pulsed CO_2 laser radiation (pulse energy of 1 J, pulse time of 70 ns) have been investigated. The samples was irradiated by CO_2 laser in two regimes – single-mode (fluency of 40 J/cm^2) and multi-mode (fluency of 48 J/cm^2) and with two laser frequency – 975 cm^{-1} and 1076 cm^{-1}.

The images of laser spots by means of the high resolution optical microscop and atomic force microscop have been made. It has been observed that by laser action on the surface of samples two kind of structures have appeared – periodical micron-sized structures with the period length close to wave length of CO_2 laser irradiation and nanoclusters with size close to 50-100 nanometers. It has been observed that the relief depth of the periodic structures depends on laser pulses number.The maximal highness of nanoclusters at the resonant frequency of laser (1076 cm^{-1}) has been observed.

The infrared (IR) reflection spectra and luminescence spectra of irradiated surface also have been recorded. The IR reflection spectra for irradiated samples show the enhancement

* Corresponding author: anel-mggu@mail.ru

of reflectance in the region of laser frequency with the band width of 20 cm^{-1} for crystal quartz. More high enhancements have been found at the frequency of 1076 cm^{-1}. It has been established that the reflectance enhancement has an accumulating character, namely, dependence of this ones on the number of laser pulses incident upon the samples take place. The luminescence spectra of irradiated sample have shown the lines in region of 380 nm for fused quartz and the line with two maximum - at 415nm and 440 nm - for crystal quartz. We believe that these spectra belong to silicon nanocrystals as long as these ones are similar to the luminescence spectra of silicon nanocrystals produced by other ways. It has been known that such spectra is associated with silicon nanocrystals having size from 2 nm to 10 nm [1, 2]. We suggest that the 50 – 100 nm sized nanoclusters mentioned above represent the aggregations of a silicon 2 – 5 nm sized nanoclusters .

We can made some conclusion:

- The periodical micro-and nano-structures in a crystal and amorphous silica have appeared because of interference of incident waves and surface waves induced by incident waves in resonant absorption media. That structures arise because of the increase of ablation velocity at the maxims of standing waves.
- Cumulative properties of periodical structures formation show us that the mainly ablation processes take place by laser action on silica surface.
- Dependence of relief depth from frequency of laser action shows that microstructure formation connects with the resonant interaction of CO_2 laser radiation with silicates. Intensive ablation in maxims of standing waves connects with the breaking of covalent oxygen – silicon bonds in region of laser action frequency what can be proved by IR spectra of irradiated samples [3].

References

1. V. Svrcek, T. Sasaki, Y. Shimizi, N. Koshisaki, Appl. Phys. Lett. **89**, 213113 (2006)
2. S. Yang, W. Cai, H. Zeng, Z. Li, J. Appl. Phys. **104**, 023516 (2008)
3. A. Mukhamedgalieva, A. Bondar, Proc. SPIE **2118**, 224 (1994)

First principle calculation of structure and lattice dynamics of $Lu_2Si_2O_7$

D.V. Nazipov[1,*], and A.E. Nikiforov[1]

[1]Ural Federal University, 620002 Ekaterinburg, Russia

Abstract. Ab initio calculations of crystal structure and Raman spectra has been performed for single crystal of lutetium pyrosilicate $Lu_2Si_2O_7$. The types of fundamental vibrations, their frequencies and intensities in the Raman spectrum has been obtained for two polarizations. Calculations were made in the framework of density functional theory (DFT) with hybrid functionals. The isotopic substitution was calculated for all inequivalent ions in cell. The results in a good agreement with experimental data.

In this work we present ab initio calculations of the crystal structure and phonon spectra of the lutetium pyrosilicate. Lutetium pyrosilicate (LPS) is a good optical matrix [1-3]. Doped with rare-earth ions LPS actively studied as a perspective laser material. LPS crystal have a low symmetrical monoclinic crystal structure, space group C2/m. Low symmetrical crystals are hard for investigation from experimental point of view. In the present study, crystal structure parameters and vibrational spectra obtained in good agreement with experiment.

Investigations performed in molecular orbitals LCAO approximation, using DFT method with hybrid functional WC1LYP in program package CRYSTAL14 [4]. Ions was described by full-electron basis sets (for Si^{3+}, O^{2-}) and by pseudopotential for Bi^{3+}.

In the work [5], devoted to a detailed study of the vibrational Raman spectrum of lutetium pyrosilicate, first identified the complete set of lattice vibrations of LPS and pyroion $[Si_2O_7]^{6-}$.

Challenging task from experimental point of view was identifying close lying lines such as № 6, 7 and № 14, 15. Calculations are performed for an ideal crystal and without the influence of the birefringence effect which makes difficult to separate A_g and B_g modes in the spectrum. Also hard to identify low intensity modes. In the work [5] was identified № 3 mode, which was absent in the work [6]. According to our results the № 3 mode has a 300 times smaller intensity than most intense mode № 8. Obtained spectra have been analyzed using isotopic shift method and ions involved in vibrations have been identified. According to our data, oxygen O1, which forms a bridge bound Si-O1-Si, involved only in high-frequent A_g=927 cm^{-1} mode. Low frequency modes correspond to vibrations of Lu heavy ion. Ions of Si have biggest activity predominantly in modes with frequency more than 490 cm^{-1}. Summing up, the identification of full set of Raman active phonons, which obtained in [5] for the first time, is confirmed by first principle DFT calculations and spectrum is analyzed by the isotopic substitution method.

* Corresponding author: qwear0@gmail.com

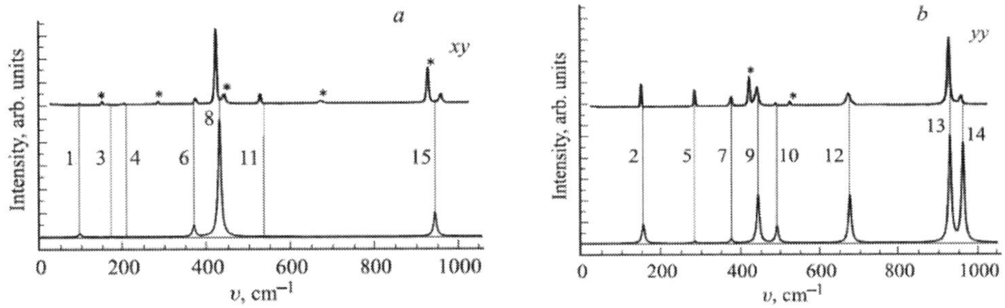

Fig. 1. Comparison of experimental [5] (top) and calculated in this work (bottom) Raman spectra, in xy (a) and yy (b) polarizations.

Table 1. Frequencies of phonon modes that are active in the Raman spectrum of $Lu_2Si_2O_7$ and ions involved in vibrations, determined by the method of isotopic substitution. (Experimental data are given in brackets [5])

№	Irreducible representation	Freqency, cm^{-1}	Involved ions
1	B_g	95(95)	Lu
2	A_g	157(150)	Lu
3	B_g	171(163)	Lu
4	B_g	207(204)	O3
5	A_g	285(285)	O2, O3
6	B_g	367(373)	O2
7	A_g	376(376)	O2
8	B_g	428(422)	O3
9	A_g	444(441)	O3
10	A_g	493(490)	Si, O3
11	B_g	528(527)	O2
12	A_g	675(672)	Si
13	A_g	931(927)	Si, O1, O3
14	A_g	963(955)	Si, O2
15	B_g	941(959)	Si, O3

The full set of modes of the phonon spectrum of $Lu_2Si_2O_7$ and their frequencies and types have been calculated (table 1). Intensities of modes, active in the Raman spectra, have been obtained for two polarizations xy and yy (fig. 1).

This work was supported by Ministry of Education and Science of the Russian Federation (project № 3.571.2014/K), Government of Russian Federation (Contract № 02.A03.21.0006, decree № 211) and Russian Foundation for Basic Research (project № 16-33-00437-mol_a.).

References

1. D. Pauwels et al., IEEE Transactions on Nuclear Science **47**, 1787 (2000)
2. H. Yang, G. Lui, Q. Lu, Q. Wei, J. Zhuang, Q. Liu, Appl. Surf. Sci. **301**, 323 (2014)
3. E. Dell'Orto, M. Fasoli, G. Ren, A. Vedda, J. Phys. Chem. C **117**, 20201 (2013)
4. http://www.crystal.unito.it
5. U.K. Voronko, A.A. Sobol, V.E. Shukshin, I.V. Gerasimov, Fizika tverdogo tela **57**, 1398 (2015)
6. F. Bretheau-Raynal, J.P. Dalbiez, M. Drifford, B. Blanzat, J. Raman Spectr. **8**, 39 (1979)

The spectroscopy of singlets and triplets excites electronic states, spatial and electronic structure of hydrocarbons and quantum classifications in chemmotology

A.E. Obukhov*

25th State Research Institute of Chemmotology of the Ministry of Defense, 121467 Moscow, Russia

Abstract. In this work we demonstrate the physical foundations of the spectroscopy of the grounds states: E- and X-ray, (RR) Raman scattering the NMR ^{1}H and ^{13}C and IR-, EPR- absorption and the singlets and triplets electronic excited states in the multinuclear hydrocarbons in chemmotology. The parameters of UV-absorption, RR-Raman scattering of light, the fluorescence and the phosphorescence and day-lasers at the pumping laser and lamp, OLEDs and OTETs- are measurements. The spectral-energy properties are briefly studied. The quantum-chemical LCAO-MO SCF expanded-CI PPP/S and INDO/S methods in the electronic and spatial structure hidrocarbons are considered.

In this work the modern views on the nature of the conversion processes are outlined, and the variation of their rate with the type of transition and with the energy spectrum and structure of the molecule are analyzed. The relationship between spectra characteristics and phenomena of the scattering, absorption and reabsorptions, luminescence, emissions and generation and the structure of hydrocarbons molecules are discussed theoretically. The full spectrums of excited singlets and triplets electronic states (STElExSt) of the quantum-chemical LCAO-MO SCF-CI INDO/S methods were calculated in [1] (Fig. 1). Under deactivations energy of pumping in STElExSt it is transformed to optics (fluorescence (Fl), phosphorescence (Ph)) and nonradiative transitions. The spin-orbitals interactions involved in STElExSt and the effect of spatial and electronic structure and other factors are considered. The hydrocarbons are given into the five types according to their physics of phenomena and a justification of this classification. The absolute quantum yield of fluorescence (QyFl) and phosphorescence (QyPh), and the rate constant (Rc) intercombinations (IntCc) and Rc innercombinations convertions (InCc) between the lowest and highly STElExSt are offered as

$$\gamma_{fl} = k_{S_1^* \to S_0} / \left(k_{S_1^* \to S_0} + k_{S_i^* \Rightarrow S_0} + \sum_{i=1, j=1}^{n,m} k_{S_i^* \Leftrightarrow T_j} \right), \tag{1}$$

$$\gamma_{ph} = k_{T_1 \to S_0} \sum_{i=1, j=1}^{n,m} k_{S_i^* \Leftrightarrow T_j} / \left(k_{S_1^* -> S_0} + k_{S_1^* \Rightarrow S_0} + \sum_{i=1}^{n} k_{S_1^* \Rightarrow T_m} \right) \left(k_{T_1 -> S_0} + k_{T_1 \Rightarrow S_0} + k^*_{(T_1 \to T_j)} \right), \tag{2}$$

* Corresponding author: aobukhov@fo.gpi.ru

where $k_{S_1^* \to S_0}$ and $k_{T_1 -> S_0}$ are the RcFl and RcPh, respectively, $\sum_{i=1,j=1}^{n,m} k_{S_i^* <=> T_j}$ is sum the rate constants (RC) of ItCc between all types of STElExSt, $k^{*(T_1 \to T_j)}$ is the new entered empirical constant, which on efficiency of photoionization ($\boldsymbol{I}_1$*), $k_{S_1^* \Rightarrow S_0}$ and $k_{T_1 \Rightarrow S_0}$ is the RC InC and RC ItCc from lowest the STElExSt.

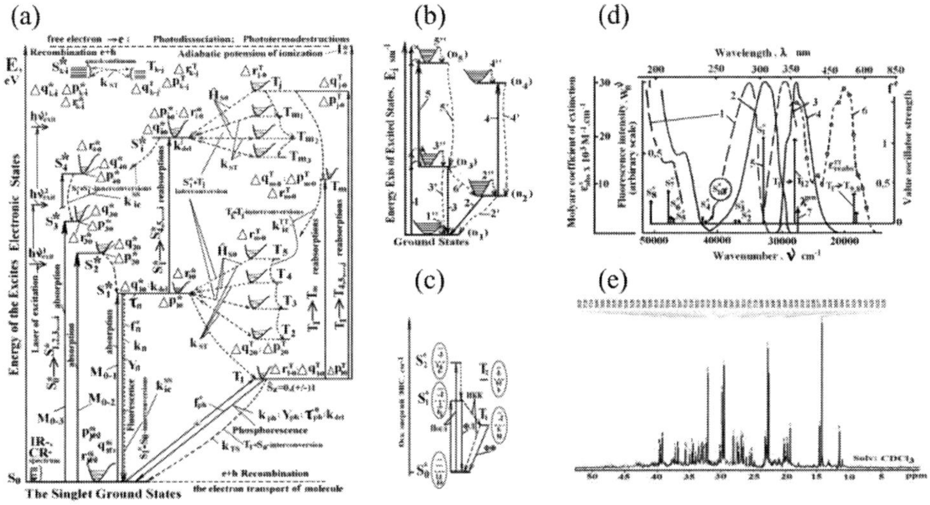

Fig. 1. (a). The full spectra of the optical and nonradiated transition between of the systems of the GrSt and between the singlets ($S^*_{1,...,i}$) and triplets ($T_{1,...,j}$) ElExSt (the ion is the quartets (Q_n) and the duplets D_m) with the LCAO-MO SCF expansion-CI INDO/S methods of calculations. According to the rules of selection on the orbital symmetry on the S^*_1->$S^*_{2,...,i}$ of singlet-singlets transitions. At the ItCC ($S^*_{1,...,i}$<≈>T_f^a) with intermediate the height triplets T_j states from the RC of ItCC is $k_{ST} \approx (10^8 \div 10^{13})$ s^{-1} effectively population the lowest S^*_1 and T_1 ElExSt. According to the rules of selection on the orbital symmetry on the T_1->$T_{1,...,j}$ of triplet-triplets transitions by populations the height T_j ElExSt it is impossible to identify with before ionizations the quasy-continuum of Rydberg states. The discrete mechanism of the multistage photoionization and photo-termo-destructions of hydrocarbons. Here $\Delta r_{\mu\nu}$ and $\Delta p_{\mu\nu}$, $\Delta q_{\mu\nu}$, and $\Delta \rho_{\mu\nu}$ – length and the stretching vibrations of bonds, change and full electronic density from atoms; (b) The five-levels of system the ElExSt. *1* is the energy of lasers pumping, *2* and *2'* is the phosphorescence (*ph*) and nonradiated transitions, *3* is of the fluorescence (*fl*) and *3'* nonradiated transitions, *4* is the singlet-singlet and *5* is triplet-triplet reabsorption (TTR) of electron-vibrations transitions (*2',3',4'* and *5'*) the InCC of transitions, (*1'',2'',3'',4'',5''*) is the electron-vibrations conversion of transitions (picosecond relaxation), *6* is the InCC of transitions. Here ($n = n_1 + n_2 + n_3 + n_4 + n_5$); (c) tree-levels of systems Yablonskii; (d) The spectra of UV-absorption (ε^{ν}_{abs},1,2) and the fluorescence (W^{ν}_{fl},3,4), and maximum the generation radiation of light (λ^{max}_{gen},7), and induced pumping the spectra of TTR (ε^{TT}_{reabs},5,6) in the vapor (2,3,5) and the solutions (1,4,6,7). Vertical continuous lines is values of oscillator strength ($f^e_{fl,phosf}$ and f^e_{reabs}) on frequencies (ν^{ij}) of the electronic S^*_0->$S^*_{1,...,i}$, and T_1->$T_{1,...,j}$ transitions $\pi\pi^*$- and $n\pi^*$-types and the values of designed by the LCAO-MO SCF-CI INDO/S for the lasers-active 2-phenyl-1,3-oxazole (POP); (e) – dope fuel of Haitec-580.

Unusual behavior of the dysprosium CF energies in Dy_2BaNiO was detected in the temperature range lower than T_N. All CF levels experience a shift (up to 10 cm^{-1}) from their positions at T_N (see, e.g., Fig. 1). Such behavior can be explained as being due to the spontaneous electric polarization that appears in the temperature range discussed [8]. Redistribution of the charge density influences the crystal field acting on dysprosium and leads to a renormalization of CF self-energies.

References

1. A.E. Obukhov, *Spectroscopy of the ground and excited states of multinuclear compounds in different conditions* (Satellite+, Moscow, 2010)

Singlet oxygen luminescence detecting in presence of hybrid associates of colloidal Ag_2S quantum dots with methylene blue molecules

O.V. Ovchinnikov[1,*], T.S. Kondratenko[1], M.S. Smirnov[1], A.S. Perepelitsa[1], I.G. Grevtseva[1], Y.A. Vinokur[1], S.V. Aslanov[1], and A.S. Matsukovich[2]

[1]Voronezh State University, 394018 Voronezh, Russia
[2]B.I. Stepanov Institute of Physics, National Academy of Sciences of Belarus, 220072 Minsk, Belarus

Abstract. In our work we demonstrate some spectroscopic investigation of colloidal Ag_2S QDs associates with methylene blue. The photosensitizing of singlet oxygen by associates of colloidal Ag_2S QDs with methylene blue was found.

Colloidal Ag_2S quantum dots (QDs) are promising material for fluorescent labeling of biological objects, including tissues affected by severe diseases. It is possible due to their intense IR luminescence. Luminescence in the region of 800-1200 nm corresponds to the therapeutic window of transparency of biological objects [1-3]. Hybrid association of Ag_2S QDs with molecules of thiazine dyes opens additional possibilities to control the quantum yield of QDs luminescence [3], and also to photosensitize singlet oxygen ($^1O_2$). This is interesting for photodynamic therapy, particularly in combination with high-contrast fluorescent marking [4,5]. Furthermore, conjugation of QDs with dye molecules can to provide higher stability of dye, for example, methylene blue (MB) in the form, which produces singlet oxygen [6].

There is a problem of unambiguous fluorescent detection of singlet oxygen in the region of 1270 nm due to Ag_2S QDs luminescence with peak at 1200 nm for most of known methods of their synthesis. This problem is also deepened due to significant duration of luminescence decay of Ag_2S QDs, which is overlapped with kinetics of luminescence of singlet oxygen. Therefore we solved the problem of changing the position of IR luminescence peak of Ag_2S QDs. This peak has weak size dependence. Its position was changed by replacing of stabilizer. Colloidal Ag_2S QDs were synthesized in aqueous gelatin solution and using thioglycolic acid (TGA).

The possibility of singlet oxygen photosensitization in water-ethanol solutions of mixtures of colloidal semiconductor Ag_2S QDs, conjugated with MB molecules was investigated for synthesized samples. We used colloidal semiconductor Ag_2S QDs with average size of 2.2 – 3.0 nm and MB molecules with concentration, provided QDs:MB=1:3 and QDs:MB=1:8 ratio respectively.

Detection of singlet oxygen production was realized using a technique, based on the measurement of $^1O_2$ molecules phosphorescence at 1270 nm. Registration of $^1O_2$ molecules phosphorescence was realized using a highly stable low-noise photodiode PDF-10C/M and

* Corresponding author: ovchinnikov_o_v@rambler.ru

monochromator MDR-4. The excitation sources was laser diode LPC-836 (λ_{max} = 660 nm, P_{max} = 250 mW).

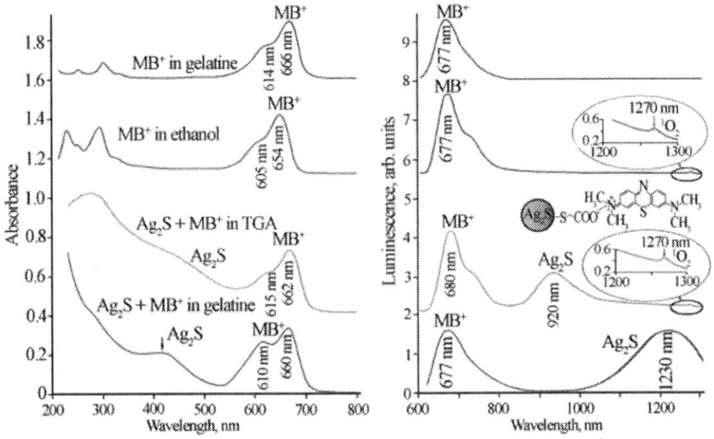

a b

Fig. 1. UV-Vis absorption (a) and luminescence (b) spectra of MB in deferent solvents and mixtures of colloidal Ag_2S QDs with MB in gelatin and TGA. Insert is scheme of interaction between Ag_2S QDs and MB in TGA.

As result of investigations it was found: 1) Synthesized Ag_2S QDs in gelatin have an average size of 2.2 nm (according TEM) and 3.0 nm in TGA. The size effect is observed in UV-Vis absorption spectra (fig. 1, a). Replacement of stabilizer from gelatin to TGA provides a shift of Ag_2S QDs luminescence peak from 1230 nm to 920 nm for similar size of Ag_2S QDs (fig. 1b). This result indicates interface nature of center of recombinational IR luminescence of Ag_2S QDs. 2) A slight shift of peak of the order 4-5 nm is observed in UV-Vis absorption spectra of MB molecules during formation of mixtures with Ag_2S QDs. This indicates a relatively weak interaction between MB molecules and Ag_2S QDs, realized probably by dipole-dipole mechanism with involving of dimethylamino groups of MB and incomplete bonds of QDs and also stabilizer molecules (fig. 1). 3) The association leads to a redistribution of the luminescence intensity of QDs and MB molecules. In the case of mixtures of Ag_2S QDs and MB in TGA and gelatin we occur quenching of QDs luminescence under excitation from the absorption band of MB. Observed regularities are manifestation of interaction between MB molecules and Ag_2S QDs. This fact can be explained by ionization of Ag_2S QDs due to transfer of photoexcited electron to MB^+, owned acceptor properties in its triplet state. 4) Singlet oxygen photosensitization was found in ethanol solution of MB (fig 1b). The peak of singlet oxygen phosphorescence at 1270 nm was detected in photoluminescence spectra of mixtures of Ag_2S QDs and MB in TGA [5]. It is not typical for photoluminescence spectrum of pure Ag_2S QDs in TGA. To observed similar peak for mixtures Ag_2S QDs and MB in gelatin is difficult due to intense luminescence of Ag_2S QDs (1230 nm).

The work is supported by Russian Foundation of Basic Research (project №16-32-00503-mol_a).

References

1. G. Honge et al., Angew. Chem. **124**, 9956 (2012)
2. I. Hocaoglu et al., J. Mater. Chem. **22**, 14674 (2012)
3. T.S. Kondratenko et al., Technical Physics Letters **42**, 365 (2016)
4. A. Fernandez-Fernandez et al., Appl. Biochem. Biotechnol. **165**, 1628 (2011)
5. S.Yu. Egorov, A.A. Krasnovsky, Biophysika. **28**, 497 (1983)
6. L. Li, Chem. Commun. **51**, 14357 (2015)

EPJ Web of Conferences **132**, 03039 (2017)
SPECTROSCOPY.SU 2016 DOI: 10.1051/epjconf/201713203039

Crystal structure and vibrational properties of $RFe_3(BO_3)_4$ (R = Ce – Lu) ferroborate crystal: *ab initio* calculations

V.P. Petrov[1,*], V.A. Chernyshev[1], and A.E. Nikiforov[1]

[1]Ural Federal University, 620002 Ekaterinburg, Russia

Abstract. The *ab initio* calculations of the crystal structure and lattice dynamics of ferroborate crystal family $RFe_3(BO_3)_4$ (R = Ce, Pm, Eu, Gd, Tb, Dy, Ho, Er, Tm, Yb, Lu), S.G. $R32$, has been carried out within the framework of the MO LCAO approach by using density functional theory and effective 4*f*-in-core pseudopotential for rare earth element. The fully optimized geometry as well as vibrational frequencies has been calculated.

We calculated the crystal structure of $RFe_3(BO_3)_4$ (S.G. R32), and afterward the phonon spectrum was calculated. The calculations were performed in the framework of the density functional theory by using hybrid functional B3LYP. The program CRYSTAL14 [1] intended for periodic calculations in the MO LCAO approach was used. ECP ("effective 4*f*-in-core pseudopotential") was used for the rare earth ions [2]. This approach has shown good results and the agreement with experiment for the crystals $RFe_3(BO_3)_4$ (R = Pr, Nd, Sm) [3]. In this work we carried out similar calculations for the remaining ions of the rare earth row. The results of calculations of the crystal structure are given in Table 1. Axis z is directed along C_3 and x along C_2. The crystal structure consist of a spiral chains of "octahedral" FeO_6 placed around the trigonal C_3 axis and straight chains of RO_6 prisms strung on the trigonal axis. The structure contains two types of BO_3 triangles also. One type is equilateral triangles BO_3 strung on the trigonal axis between RO_6 prisms; another type is isosceles triangles that are placed aside of the trigonal axis and connect chains RO_6 and FeO_6. The calculations predict that in the middle of the row (R=Dy, Ho) the isoscele triangle $B2O_3$ becomes almost equilateral. The lattice parameters a and c decrease in the row La-Lu that corresponds to the lanthanide contraction. The distorted octahedron FeO_6 in the end of the series (R=Lu) is closer to the ideal octahedron.

Table 1. Lattice parameters (a, c) and bond lengths (Å) of structure-forming units of $RFe_3(BO_3)_4$

R	a	c	RO_6 prism	FeO_6 "octahedron"			$B1O_3$ equilateral triangle	$B2O_3$ isosceles triangle	
Ce	9.6986	7.7512	2.467	1.972	2.064	2.091	1.388	1.371	1.386
Pm	9.6682	7.7026	2.431	1.980	2.055	2.077	1.385	1.373	1.383
Eu	9.6500	7.6778	2.410	1.984	2.050	2.070	1.383	1.374	1.380
Gd	9.6409	7.6674	2.401	1.986	2.048	2.066	1.382	1.374	1.379
Tb	9.6306	7.6551	2.389	1.989	2.045	2.063	1.381	1.375	1.378
Dy	9.6223	7.6454	2.380	1.991	2.043	2.060	1.380	1.376	1.377
Ho	9.6142	7.6361	2.371	1.993	2.041	2.056	1.379	1.376	1.377
Er	9.6059	7.6270	2.363	1.995	2.039	2.054	1.378	1.375	1.377

* Corresponding author: lancervlad@gmail.com

Tm	9.5985	7.6191	2.355	1.996	2.037	2.051	1.377	1.374	1.378
Yb	9.5908	7.6116	2.347	1.998	2.035	2.048	1.377	1.373	1.378
Lu	9.5843	7.6047	2.340	2.000	2.034	2.046	1.376	1.372	1.379

Table 2. Vibrational frequencies, cm^{-1} of $RFe_3(BO_3)_4$ (R = Ce–Lu).

Irrep.	Ce	Pm	Eu	Gd	Tb	Dy	Ho	Er	Tm	Yb	Lu
A_1	181	182	182	183	183	184	184	184	184	184	185
	291	299	302	305	307	309	312	314	316	318	319
	463	465	463	466	467	468	469	469	470	471	472
	644	648	649	649	650	651	652	652	652	653	653
	941	950	955	957	959	960	961	962	962	962	962
	990	989	988	988	987	988	988	989	991	992	993
	1202	1212	1219	1223	1227	1230	1234	1237	1241	1244	1247
A_2	34	43	46	50	53	55	59	60	62	67	65
	155	159	160	161	161	161	162	162	162	161	163
	191	195	193	194	193	191	190	187	185	183	181
	206	204	202	203	204	205	207	208	209	209	211
	258	258	256	257	256	256	255	255	255	252	254
	310	300	289	288	282	279	276	273	270	267	265
	337	339	341	341	343	344	345	347	348	352	350
	370	376	380	382	384	385	387	388	390	392	393
	620	621	621	621	622	622	622	623	623	623	623
	694	690	690	691	691	691	691	690	690	690	690
	755	751	749	747	746	745	744	743	743	759	741
	1268	1265	1262	1261	1258	1257	1256	1254	1253	1251	1250
E	86	87	87	87	87	87	87	87	87	87	88
	160	161	161	162	161	161	161	161	161	161	161
	192	196	198	199	200	201	203	204	204	205	206
	239	238	234	236	235	235	235	234	234	233	233
	265	269	269	271	271	272	272	272	272	272	272
	273	275	276	278	280	280	282	283	283	283	283
	316	316	316	317	317	317	317	318	318	318	319
	350	351	346	345	342	339	337	336	334	331	330
	363	366	368	371	372	374	375	377	378	379	380
	377	379	379	382	383	384	386	387	389	390	391
	430	431	432	433	434	435	437	438	439	440	441
	577	580	581	581	581	582	582	583	583	583	583
	619	623	625	626	627	628	630	631	632	633	634
	662	665	667	668	668	669	670	671	672	672	673
	754	751	752	752	752	752	752	753	753	753	753
	968	969	968	968	968	967	967	967	966	966	965
	1159	1174	1183	1187	1192	1195	1198	1203	1206	1209	1212
	1195	1203	1209	1213	1217	1220	1222	1226	1229	1232	1234
	1293	1289	1286	1284	1281	1280	1279	1278	1277	1276	1275

The frequencies of normal optical vibrations $\Gamma = 7A_1 + 12A_2 + 19E$ are given in Table 2. According to the calculations, by replacement of the rare earth ion, the lowest A_2 mode has the most significant changes. In this mode, the translations of the rare earth ion [3] are the most significant. High-frequency phonons, which are caused by BO_3 groups, react less along the Ce-Lu row. These are modes: A_1 in range 640-1200, A_2 in range 620-1270, E in range 570-1290 cm^{-1}. The degree of participation of the ions in the modes was analyzed in the previous study [3]. It should be noted that the using of *ab initio* approach allows us to describe the structure and lattice dynamics of $RFe_3(BO_3)_4$ for the all R=Ce-Lu.

The support from the Ministry of Education and Science (pr. no. 3.571.2014/K), the Government of RF (contr. no. 02.A03.21.0006, stat. no. 211), the Russian Foundation for Basic Research (V.P.P., pr. no. 16-33-00437-mol_a) is acknowledged.

References

1. R. Dovesi et al., Int. J. Quantum Chem. **114**, 1287 (2014)
2. http://www.tc.unikoeln.de/PP/clickpse.en.html
3. V.A. Chernyshev et al., Phys. of the Solid State **58**, 1642 (2016)

EPJ Web of Conferences **132**, 03040 (2017) DOI: 10.1051/epjconf/201713203040
SPECTROSCOPY.SU 2016

Ultrafast dynamics of photoprocesses induced by femtosecond IR laser radiation in iron pentacarbonyl molecules and clusters

D.G. Poydashev*, V.O. Kompanets, V.N. Lokhman, S.V. Chekalin, and E.A. Ryabov

Institute for Spectroscopy RAS, 108840 Troitsk, Moscow, Russia

Abstract. Ultrafast dynamics of photoinduced processes in $[Fe(CO)_5]_n$ clusters and free $Fe(CO)_5$ molecules excited by resonant femtosecond IR radiation ($\lambda \sim 5$ μm) were studied in a molecular beam. Using a pump-probe teqnique characteristic rates of these processes were obtained and theoretically analyzed.

The interaction of resonant femtosecond IR radiation with CO vibrational modes of both free $Fe(CO)_5$ molecules and $[Fe(CO)_5]_n$ molecular clusters followed by subsequent photoinduced processes are the main subject of the studies presented in this report. The experiments were carried out using a combination of femtosecond laser spectroscopy and photoionization time-of-flight mass spectrometry (TOF MS) techniques. The clusters and cold free molecules were produced by supersonic expansion of $Fe(CO)_5$ + Ar mixture from a pulsed nozzle.

The technique of the multiphoton ionization of molecules by femtosecond ultraviolet or visible radiation in combination with time-of-flight mass spectrometry has been used for the first time to probe the dynamics of processes in vibrationally excited molecules in a molecular beam. It has been found that the resonant excitation of vibrations of CO bonds in free $Fe(CO)_5$ molecules below their dissociation threshold results in an "instantaneous" (on a time scale of ≈ 800 fs) significant decrease in the signal from the $Fe(CO)_5^+$ molecular ion (see Fig. 1). The subsequent time relaxation of this signal, as well as the dependence of an induced change in this signal on the energy fluence Φ_{IR} of the infrared pump pulse, has been studied.

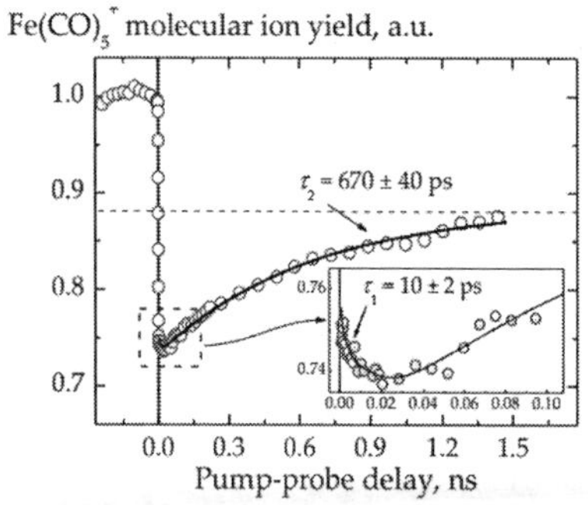

Fig. 1. Dynamics of $Fe(CO)_5^+$ molecular ion yield in the case of IR excitation of free molecules.

* Corresponding author: poydashev@isan.troitsk.ru

The analysis of the results indicates that the infrared pulse selectively excites vibrational modes of the CO bonds. A decrease in the multiphoton ionization cross section for the probe pulse as a result of this excitation leads to an initial decrease in the yield of $Fe(CO)_5^+$ molecular ion . It has been found that this signal further relaxes with two characteristic times – short, $\tau_1 \approx 10$ ps, and long, $\tau_2 \approx 670$ ps (see Fig. 1). Slow relaxation is due to the intramolecular relaxation of the energy localized in the vibrations of CO bonds to other vibrations of the molecule, and τ_2 is the characteristic time of this process. The time τ_1 is possibly associated with Berry pseudorotation (for details, see ref.1).

Using the femtosecond time-resolved infrared pump – visible probe technique, we have measured for the first time the ultrafast dissociation dynamics of $[Fe(CO)_5]_n$ clusters induced by IR resonant excitation of CO vibrations in the 5μm region. Relaxation of vibrational energy stored by clusterized molecules leads to an increase of the temperature of the cluster and subsequent evaporation of free $Fe(CO)_5$ molecules formed as a result of the cluster dissociation. These molecules have been ionized by the femtosecond laser radiation at λ = 400 nm and have been detected by TOF MS device (left side of Fig. 2).

The temporal dependence of the yield of free molecules has been measured under different conditions of the IR laser excitation. Experiments with femtosecond IR radiation with different widths of the spectrum (100 cm^{-1} and 185 cm^{-1}) have allowed us to conclude that the dynamics of the yield of free molecules from clusters is determined by the magnitude of the energy absorbed by the cluster, with the molecules in the cluster utilizing only the «resonant» part of the radiation spectrum.

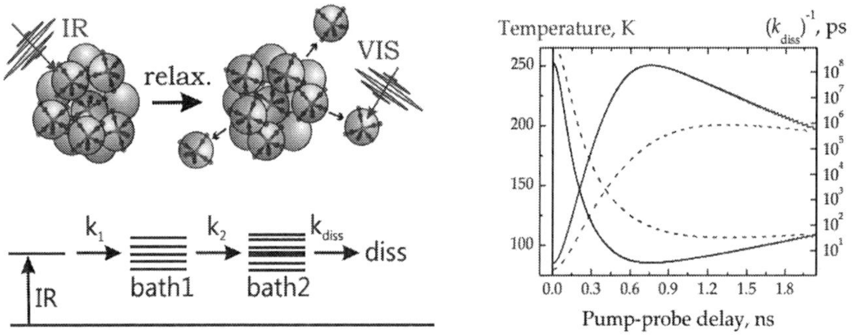

Fig. 2. Left: schematic diagram of the processes involved in the dissociation of clusters. Right: dynamic dependencies for cluster temperature T and dissociation rate k_{diss} in $[Fe(CO)_5]_n$ clusters. Solid and dashed lines are calculations for 100 cm^{-1} and 185 cm^{-1} IR bandwidth respectively.

We have proposed a model that describes well experimental results and makes it possible to calculate the temporal profile of the cluster temperature as well as cluster dissociation rate k_{diss} in terms of the concept of the evaporative ensemble (see right side of Fig. 2). The rates of the *intramolecular* and *intracluster* vibrational relaxation in $[Fe(CO)_5]_n$ clusters have been estimated (for the details of the modelling see ref. 2). For instance, the model shows that the characteristic time of the initial energy relaxation from excited CO vibrations to low-frequency modes of the *molecule* in cluster is close to the characteristic time of the subsequent vibrational energy redistribution from low-frequency modes of the molecule into the intermolecular vibrations of the cluster and equals to approx. 280 ps.

References

1. V.O. Kompanets, V.N. Lokhman, D.G. Poydashev, S.V. Chekalin, E.A. Ryabov, JETP **122**, 621 (2016)
2. D.G. Poydashev, V.N. Lokhman, V.O. Kompanets, S.V. Chekalin, E.A. Ryabov, J. Phys. Chem. A **118**, 11177 (2014)

High-resolution spectroscopy, crystal-field calculations, and quadrupole helix chirality of $DyFe_3(BO_3)_4$

M.N. Popova[1,*], B.Z. Malkin[2], T.N. Stanislavchuk[3], E.P. Chukalina[1], K.N. Boldyrev[1], and I.A. Gudim[4]

[1]Institute for Spectroscopy RAS, 142190 Troitsk, Moscow, Russia
[2]Kazan Federal University, 420008 Kazan, Russia
[3]Department of Physics, New Jersey Institute of Technology, NJ 07102 Newyark, USA
[4]Kirensky Institute of Physics, Siberian Branch of RAS, 660036 Krasnoyarsk, Russia

Abstract. High-resolution polarized transmission spectra of $DyFe_3(BO_3)_4$ single crystals were investigated in broad spectral (10-23000 cm^{-1}) and temperature (3.5-300 K) ranges. Energies of the dysprosium levels in the paramagnetic and antiferromagnetic phases were determined. On the basis of these data and preliminary calculations in the frameworks of the exchange-charge model, we determined the crystal-field and Dy-Fe exchange interaction parameters of the Dy^{3+} ions at sites with the point C_2 symmetry corresponding to the enantiomorphic $P3_121$ and $P3_221$ space groups. The values of electronic quadrupole moments of the Dy^{3+} ions were calculated, which enabled us to interpret results of the work [Usui et al., Nature Mater. 13, 611 (2014)] on the observation of domains of different quadrupole chirality in $DyFe_3(BO_3)_4$.

The crystals $DyFe_3(BO_3)_4$ belong to the family of multiferroic rare-earth iron borates that crystallize in the trigonal $R32$ structure of the natural mineral huntite. With decreasing the temperature to T_S=285 K, $DyFe_3(BO_3)_4$ undergoes a structural phase transition from the $R32$ phase to the enantiomorphic space-group pair $P3_121$ and $P3_221$. It has been shown recently by resonant x-ray diffraction that, below T_S, single-crystal samples of $DyFe_3(BO_3)_4$ contain macroscopic domains which differ by right-handed ($P3_121$) or left-handed ($P3_221$) helical structures of electronic quadrupole moments of the Dy^{3+} ions [1]. Quadrupole moments at the Dy sites are induced by a low-symmetry C_2 component of the crystal field that appears below T_S and distorts the electronic density distribution in the Dy^{3+} ions. It is possible to calculate values of these moments on the base of the crystal-field (CF) parameters. A limited available information on the CF levels of Dy^{3+} ions in $DyFe_3(BO_3)_4$ prevents from doing CF calculations in the frame of the C_2 symmetry CF Hamiltonian with 15 parameters.

We have measured temperature-dependent transmission spectra of $DyFe_3(BO_3)_4$ single crystals corresponding to optical transitions from the ground multiplet to all the excited multiplets up to about 22200 cm^{-1} and to the far-infrared transitions within the ground multiplet. Almost all of the spectral lines observed in the absorption spectra of both the

* Corresponding author: popova@isan.troitsk.ru

paramagnetic and antiferromagnetic (T< T_N=39 K) phases of $DyFe_3(BO_3)_4$ were unambiguously identified and the corresponding energies of excitations of the Dy^{3+} ions were determined. This dataset formed a basis for the CF calculations and modeling of exchange interactions. Here, we confine ourselves with the paramagnetic phase.

The spectrum of a Dy^{3+} ion in a dielectric crystal can be described using the Hamiltonian $H=H_{FI}+H_{CF}$, where H_{FI} is the free-ion Hamiltonian but $H_{CF}=\sum_{p,q} C_q^p B_q^p$ is the CF Hamiltonian determined by 15 independent CF parameters B_q^p (C_q^p are the spherical tensor operators of rank p). The initial values of the CF parameters were calculated in the framework of the exchange-charge model [2]. Then, the energies of transitions between the levels of the Dy^{3+} ions, obtained from the numerical diagonalization of the Hamiltonian H operating in the total space of 2002 states of the electronic $4f^9$ configuration, were compared with the measured optical spectra of $DyFe_3(BO_3)_4$ in the paramagnetic phase, and the initial CF parameters were varied to fit the experimental data. Values of the six CF parameters ($B_0^2=404$, $B_0^4=-1192$, $B_{-3}^4=B_3^4=554i$, $B_0^6=328$, $B_{-3}^6=B_3^6=70i$, $B_6^6=B_{-6}^6=232\,\mathrm{cm}^{-1}$) which determine the dominant trigonal CF component change monotonously along the series of RE iron borates. The remaining nine parameters of the CF component of the C_2 symmetry ($B_{-1}^2=B_1^2=58.4i$, $B_2^2=B_{-2}^2=69.4$, $B_{-1}^4=B_1^4=-49.2i$, $B_2^4=B_{-2}^4=101.2$, $B_4^4=B_{-4}^4=15.9$, $B_{-1}^6=B_1^6=-7.4i$, $B_2^6=B_{-2}^6=-14$, $B_4^6=B_{-4}^6=31.4$, $B_{-5}^6=B_5^6=-79i$) are found with lower accuracy because they little influence the multiplet splittings. However, just this CF component determines some specific properties of RE iron borates in the $P3_121$ (or $P3_221$) phase, in particular, the anisotropy of the magnetic spectroscopic factors of Kramers doublets of the Dy^{3+} ions in the *ab*-plane and the quadrupole helix chirality in $DyFe_3(BO_3)_4$ [1].

The low-symmetry C_2-component of the crystal field distorts the electronic density distribution in the Dy^{3+} ions and induces non-zero non-axial components of the electronic quadrupole moment at the dysprosium sites. The quadrupole helix chirality appears due to rotations of the local C_2 symmetry axis by $2\pi/3$ and $4\pi/3$ at the $3a$ dysprosium sites shifted relative one another along the c-axis by $c/3$ and $2c/3$, respectively. The results of calculations of the average values of the quadrupole moment components $<Q>=\mathrm{Tr}[Q\exp(-H/k_BT)]/\mathrm{Tr}[\exp(-H/k_BT)]$ at different temperatures T (k_B is the Boltzman constant) using the set of the CF parameters given above agree satisfactorily with the experimental data presented in Ref. [1] (see Table 1) and confirm supposition [1] that changes of $<Q_{x^2-y^2}>$ and $<Q_{yz}>$ with temperature are mainly induced by redistribution of populations of CF sublevels of the Dy^{3+} ion ground multiplet.

Table 1. Quadrupole moments (arbitrary units) of Dy^{3+} in the $P3_121$ phase of $DyFe_3(BO_3)_4$.

Temperature	$<Q_{x^2-y^2}>$		$<Q_{yz}>/<Q_{x^2-y^2}>$	
	Measured [1]	Computed	Measured [1]	Computed
50 K	-1.22	-1.18	-1.23	-1.415
100 K	-0.906	-0.910	-1.20	-1.202
150 K	-0.678	-0.695	-1.10	-1.096
200 K	-0.50	-0.548	-0.90	-1.031

This paper was supported by the Russian Science Foundation (14-12-01033).

References

1. T. Usui et al., Nature Materials **13**, 611 (2014)
2. B.Z. Malkin, *Ion-phonon interactions*, *Spectroscopic Properties of Rare Earths in Optical Materials* (Springer, Berlin, 2005)

DOI: 10.1051/epjconf/201713203042

Interaction of phase-diffusion field with a molecular gas

L.S. Revin[1,2,*], V.L.Vaks[1,2], V.P. Koshelets[3], and H. Wang[4]

[1]Institute for Physics of Microstructures RAS, 603950 Nizhny Novgorod, Russia
[2]Lobachevsky State University, 603950 Nizhny Novgorod, Russia
[3]Kotel'nikov Institute of Radio Engineering and Electronics RAS, 125009 Moscow, Russia
[4]National Institute for Materials Science, Tsukuba, Japan

Abstract. You should leave 8 mm of space above the abstract and 10 mm after the abstract. The heading Abstract should be typed in bold 9-point Arial. The body of the abstract should be typed in normal 9-point Times in a single paragraph, immediately following the heading. The text should be set to 1 line spacing. The abstract should be centred across the page, indented 17 mm from the left and right page margins and justified. It should not normally exceed 200 words.

The coherent signal induces macroscopic polarization in the molecular gas at the resonant frequency [1]. Based on this molecular response concentration of studied component of the gas mixture is determined with high accuracy [2]. To obtain high resolution a source with extremely narrow generation lines is usually used. This is achieved by applying a phase-lock loop (PLL) system combined with an adjustment system, which considerably complicates the device structure.

An alternative way is to use a radiation source with a noise signal. Previously it was predicted [3] that phase-diffusion field (PDF) of the terahertz range induces in the system macroscopic polarization comparable in magnitude to the polarization in case of a coherent signal. The PDF source does not require a PLL system that simplifies the spectrometer. In this case the resolution is determined by the spectral characteristics of the receiver.

At present the only spectrometer using a noise radiation source in THz ranges is a Fourier transform spectrometer based, for example, on a heated ceramic plate. This radiation source generates almost "white" noise and requires a Michelson interferometer to obtain the molecular spectra. Spectral resolution in this case is determined by the finite path of a moving mirror which is insufficient for gas spectroscopy.

The present study is aimed to demonstrate a new method of GHz absorption spectroscopy based on the noise source of radiation.

The experiment was as follows: the source emitting phase-diffusion signal was an oscillator based on the stacks of $Bi_2Sr_2CaCu_2O_8$ (BSCCO) intrinsic Josephson junctions [4-5]. The backward-wave oscillator was used as a source of coherent signal [6]. As a molecular gas the ammonia mixed with water (10% solution) was probed. The signal was measured by the superconducting integrated receiver with local oscillator phase locked at

* Corresponding author: rls@ipmras.ru

566 GHz and intermediate frequency region 4 - 8 GHz [7-8]. As the results of measurements the absorption line at 572.5 GHz was observed.

The working area of BSCCO was in the frequency range of 400-600 GHz, the linewidth approximated by the Lorentz profile was reached 200 MHz [9[. In the case of BSCCO linewidth a hundred times larger than the linewidth of the molecular spectra, on the spectral characteristics the power drawdown was observed. This value of power was equal to the power absorbed by molecular gas. Thus the absorption coefficient of ammonia was calculated as the ratio of the absorbed power to the incident one.

In the experiment with source of coherent signal the backward-wave oscillator operating in the output range of 450 - 750 GHz was equipped with a phase-lock loop system to obtain a narrow linewidth of 20 kHz. For spectral line narrower than the linewidth of the molecular spectra it was necessary to change the frequency and register the maximum transmitted power depending on the frequency.

Thus the absorption coefficient was derived for both cases and the comparison the values obtained with two different sources were performed. Taking into account the measurement errors the obtained absorption coefficients for the case of coherent signal and phase-diffusion field was close enough. Possible reasons for the observed small deviations may be the following factors:

- In the case of phase-diffusion field there was insufficient power reached the receiver after the absorption of the gas. As a result, the signal-to-noise ratio and the absorption coefficient decreased.
- According to the theory, a noise signal in contrast to the coherent one provides general heating of the system. This decreases the difference of level populations, and, consequently, decreases the absorption coefficient.

Thus the work has investigated the possibility of using the phase-diffusion field for detection of the molecular spectra in the THz range. It is shown that the interaction of PDF with gas leads to macroscopic polarization comparable to the polarization in case of a coherent signal. Investigations demonstrate the possibility of practical application of the phase-diffusion field in the THz range high-resolution spectroscopy. In particular the BSCCO generator can be used to develop compact THz spectrometer with high resolution

The work is supported by the Ministry of Education and Science of the Russian Federation (№ 14.607.21.0100) and by Russian Science Foundation (№ 16-19-10478).

References

1. W.H. Flygare, *Molecular structure and dynamics* (Prentice-Hall, Inc., New Jersey, 1978)
2. V.L. Vaks, A.B. Brailovsky, V.V. Khodos, Infrared & Millimeter Waves **20**, 883 (1999)
3. E.A. Sobakinskaya, A.L. Pankratov, V.L. Vaks, Physics Letters A **376**, 265 (2012)
4. S. Guenon, M. Grunzweig, B. Gross et. al., Phys. Rev. B **82**, 214506 (2010)
5. R. Kleiner, Science **318**, 1254 (2007)
6. V.L. Vaks, V.V. Khodos, E.V. Spivak, Rev. Sci. Instr. **70**, 3447 (1999)
7. V.P. Koshelets, S. Shitov, Supercond. Sci. Technol. **13,** R53 (2000)
8. A.L. Pankratov, V.L. Vaks, V.P. Koshelets, J. Appl. Phys. **102**, 063912 (2007)
9. L. Mengyue, J. Yuan, N. Kinev, J. Li, B. Gross, S. Guénon, Phys. Rev. B **86**, 060505 (2012)

EPJ Web of Conferences **132**, 03043 (2017) DOI: 10.1051/epjconf/201713203043
SPECTROSCOPY.SU 2016

New laboratory spectra of the Fe^{6+} and Fe^{7+} ions

A.N. Ryabtsev*

Institute for Spectroscopy RAS, 108840 Troitsk, Moscow, Russia

Abstract. The Fe^{6+} and Fe^{7+} ion spectra excited in a vacuum spark were recorded and investigated in the 120 - 350 Å region. Previous identification is corrected and extended. A comparison with the Sun spectrum obtained from Hinode satellite was performed and identification of new lines was suggested.

The spectra of iron ions are of great value for the astrophysics because of large abundance of iron in the universe. Contemporary astronomy shows tremendous progress due to development of new instruments for ground-based as well as extra-atmospheric observations. The instruments are characterized by large spectral and spatial resolution. In particular, Extreme Ultraviolet Imaging Spectrometer (EIS) on board of the solar observatory Hinode provided unprecedented spectra of the upper transition solar region [1] containing intense Fe VII and Fe VIII lines. These spectra were studied previously in 80-th of the last century but the analyses of new solar spectra [2-4] showed that there is insistent necessity in new laboratory study of the Fe^{6+} и Fe^{7+} ion spectra.

The spectra in the vacuum ultraviolet region 120 - 350 Å were recorded at the Institute of Spectroscopy RAS on a high resolution grazing incidence spectrograph with a 3 м diffraction grating. The spectrograph has a plate factor from 0.30 through 0.46 Å/mm respectively in the 120 - 350 Å region. The instrumental width of the spectral lines was about 0.04 mm, what corresponds to 0.015 Å resolution at registration on the photoplates. A three-electrode low-inductance vacuum spark was used for the excitation of the spectra. A discharge circuit contained a capacitor C=10 or 150 mkF, charged to up to 5 kV voltage. New type of radiation-measuring instrument - the FUJI Imaging Plate- was used for spectrum recording. The spectra recorded on the Imaging Plate were scanned by Typhoon FLA 9500 scanner and digitized using the codes ImageQuant TL 7.0 and GFit , resulting in a table of line center positions and intensities. The signal linearity in the range of several orders of intensity magnitude offers a great advantage of an Image Plate in a comparison with a photographic plate. This gives an opportunity for more accurate measurement of the line intensities and their variations with a change of the excitation conditions what provided more reliable differentiation of the lines belonging to transitions in the different iron ions.

By a variation of the spark discharge conditions the iron spectra favorable for the Fe VII and Fe VIII lines were obtained. They show also differences in a behavior of the lines belonging to these spectra.

A part of the spark spectrum from the ~191 – 198 Å region recorded at two excitation conditions is represented in Fig.1 in comparison with the similar region in the solar

* Corresponding author: ryabtsev@isan.troitsk.ru

DOI: 10.1051/epjconf/201713203043

spectrum [1]. First, it is seen that the laboratory Fe VII и Fe VIII lines reproduce the solar spectrum quite well but with higher resolution, second, there is quite distinct differentiation of the lines of different stages of the ionization.

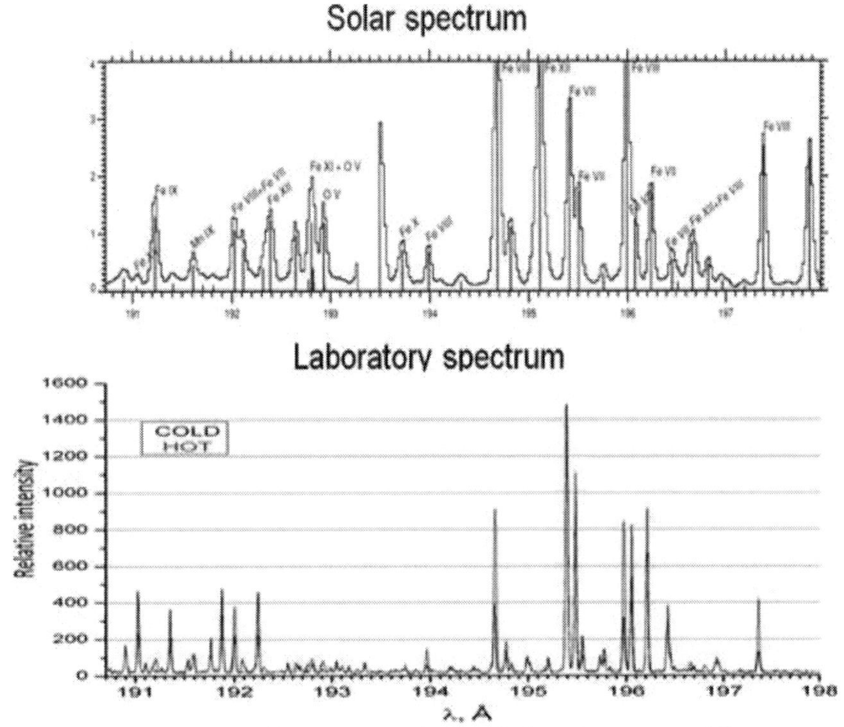

Fig. 1. Vacuum spark spectrum of iron in comparison with the solar spectrum [1].

As a result of the analysis of the Fe VII $3p^63d^2 - (3p^63d4p + 3p^53d^3)$ transitions 185 lines from the region 158 – 266 Å were identified, 55 lines being new. In Fe VII now 65 energy levels are known.

The laboratory Fe^{7+} spectrum consists now of 31 reliably identified lines.

The wavelengths and intensities measured in laboratory spectra of $Fe^{6+}$and Fe^{7+} together with the calculated transition probabilities are used for elimination of the uncertainties in the solar spectrum interpretations. It was found, that the Fe VII line identifications in the solar spectrum performed in [4] are quite questionable. Many new Fe VII lines suggested in this article are absent in our laboratory spectra, or show other properties than Fe VII or the other more intense lines from particular levels are absent. On the other hand, almost all Fe VII analysis in the solar spectrum in [2] is supported by our laboratory spectrum. New identifications are added to the solar spectrum.

The work has been performed in the framework of the research program of the RAS Physical Science Division "Basic optical spectroscopy and its applications".

References

1. P.R. Young, E. Landi, Ap. J. **706**, (2009)
2. P.R. Young, E. Landi, Ap. J. **707**, 173 (2009)
3. G. Del Zanna, A&A **508**, 501 (2009)
4. G. Del Zanna, A&A **508**, 513 (2009)

 DOI: 10.1051/epjconf/201713203044

Spectral manifestation of substitution of out-of-plane ligands in metallophtalocyanines

A. Starukhin[1,*], V. Knyukshto[1], A. Gorski[2], V. Chernii[3], I. Tretyakova[3], M. Kijak[2], and P. Kowalska[2]

[1]B.I. Stepanov Institute of Physics, NAS of Belarus, Minsk, Belarus
[2]Institute of Physical Chemistry, Polish Academy of Sciences, Warsaw, Poland
[3]V.I.Vernadskii Institute of General and Inorganic Chemistry, NAS of Ukraine, Kyiv, Ukraine

Abstract. In our study, we demonstrate the splitting effect of long-wavelength bands in absorption spectra (excitation of fluorescence) of the series of Zn(II), Ti(IV), Zr(IV) and Hf(IV) phthalocyanines at low temperatures. The effect has been explained by the shift of metal ion out of the plane of phthalocyanine macrocycle, the transformation of macrocycle to nonplanar "*dome*" conformation and additional loss the high symmetry of macrocycle.

The molecules of metalloporphyrins and metallophthalocyanines belong to the high symmetry group D_{4h} and exhibit only two wide absorption bands in the visible spectrum range at ambient temperature. The long-wavelength band corresponds to the two doubly degenerate purely electronic transitions whilst shorter - to the transitions to the vibrational sublevels of these states [1]. In the case of metalloporphyrins, the double degeneracy of electronic transitions can be removed at low temperatures due to Jahn-Teller instability or a few others factors and the energy gap between the electronic states becomes higher than 100 cm^{-1} [2-4]. In this context, it seems quite interesting to investigate the splitting effects of long-wavelength bands in the absorption spectra of metallophthalocyanines containing ligands coordinated to the central metal atom and thus located out of the macrocyclic plane and compare the data with results gained for symmetrical metallophthalocyanines and metalloporphyrins. As a reference in our studies, symmetrical metallocomplex - Zn(II) phthalocyanine was selected.The phthalocyanines containing Zn(II), Ti(IV), Zr(IV) and Hf(IV) ions were studied in the current work. The structures of Zn tetra-tret-butylphthalocyanine ($^{t\text{-}Bu}$PcZn) and complex of Zr or Hf phthalocyanine with two out-of-plane dibenzoylmethanate ligands ($PcMDbm_2$) are presented on Fig. 1. The set of absorption and fluorescence excitation experiments of phthalocyanine metallocomplexes have been carried out at low temperatures (ranging from 9 to 77 K) in 2-methyltetrahydrofuran forming an optical transparent rigid glassy matrix at these conditions.

Upon the above-mentioned In the case of $^{t\text{-}Bu}$PcZn the splitting of the long-wavelength band at 77 K was not so obvious. While, for compounds with out-of-plane ligands the

* Corresponding author: a_starukhin@mail.ru

splitting in the case of $PcZrCl_2$ has the value about 120 cm^{-1}, and reaches the maximum value (260 cm^{-1}) for $PcZrDbm_2$ metallocomplex (Fig. 2).

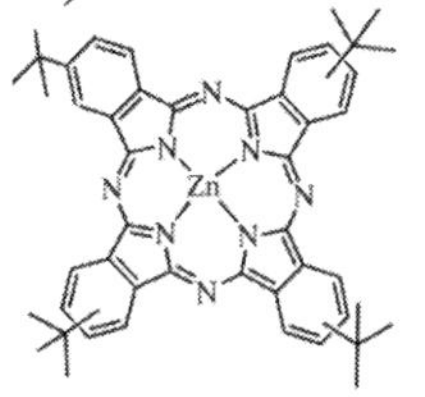
1

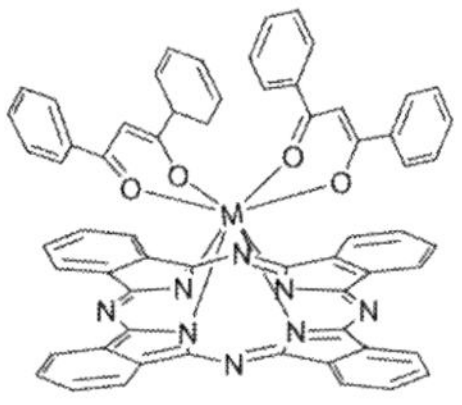
2

Fig. 1. Structure of $^{t\text{-}Bu}PcZn$ (1) and $PcMDbm_2$ (M = Zr, Hf) (2).

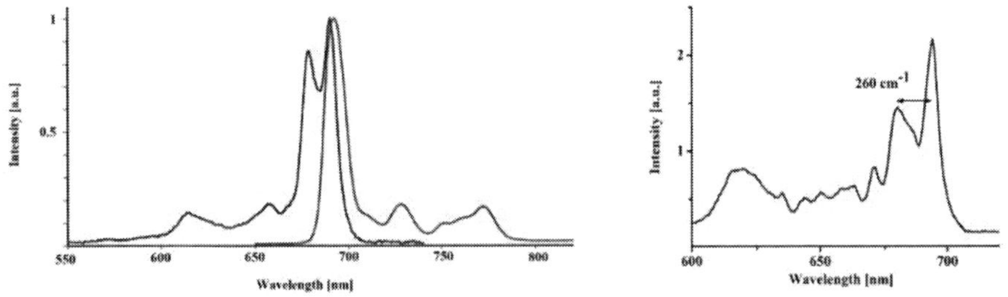

Fig. 2. Fluorescence excitation (blue) and emission (red) spectra of $PcZrDbm_2$ in 2-methyltetrahydrofurane at 77 K, left, and fluorescence excitation spectrum of $PcZrDbm_2$ at 9 K, right.

Quantum-chemical calculations, DFT methods implemented in Gaussian09 software package, were carried out for all metallocomplexes. The optimization of the molecular ground state geometry was performed using the B3LYP functional with SVP basis set. On the basis of the experimental data and molecular modeling results, we assume that the main reason for the splitting of the long-wavelength bands in the absorption spectra is the shifting of the metal ion out of the plane of the phthalocyanine macrocycle. According to calculations there is no out-of-plane shift in the case of PcZn, whereas for $PcZrCl_2$ and $PcZrDbm_2$ the shift value equals 1,03 Å and 1,24 Å, respectively. For $PcHfDbm_2$ the values of calculated shift (1,24 Å) and shift obtained from the analysis of X-ray structure (1,194 Å) [5] are well correlated and support the accuracy of performed simulations.

The splitting of long-wavelength bands in the absorption spectra of metallophtalocyanines containing out-of-plane ligands is considered to be caused by the transition of macrocyclic core to the nonplanar "dome" conformation [4,6] and additional loss by macrocycle a fourfold symmetry.

This work was supported by the European Union's Horizon 2020 research and innovation programme under grant agreement No 645628.

References

1. K.N. Solovyov, L.L. Gladkov, A.S. Starukhin, S.F. Shkirman, *The Spectroscopy of Porphyrins: Vibrational States* (Nauka i Technika, Moscow, 1985) (in Russian)
2. G. Jansen, M. Noort, Spectrochim. Acta **32**, 747 (1976)
3. A.S. Starukhin, A.M. Shulga, Opt. Spectrosc. **98**, 780 (2005)
4. V.N. Knyukshto, A.M. Shulga, E.I. Sagun, E.I. Zenkevich, J. Appl. Spectr. **65**, 900 (1998)
5. V. Chernii, I. Tretyakova, V. Bon, O. Severinovskaya, S. Volkov, Dyes Pigm. **94**, 187 (2012)
6. D.L.E. Cullen, F. Meyer, Acta Crystallographica B **32**, 2259 (1976)

EPJ Web of Conferences **132**, 03045 (2017) DOI: 10.1051/epjconf/201713203045
SPECTROSCOPY.SU 2016

Rotational spectrum of the NH_3–He van der Waals complex

L. Surin[1,*], and M. Schnell[2]

[1]Institute for Spectroscopy RAS, 108840 Troitsk, Moscow, Russia
[2]Max-Planck-Institut für Struktur und Dynamik der Materie, Hamburg, Germany

Abstract. The interaction between ammonia and helium has attracted considerable interest over many years, partly because of the observation of interstellar ammonia. The rate coefficients of NH_3–He scattering are an important ingredient for numerical modeling of astrochemical environments. Another, though quite different application in which the NH_3–He interaction can play an important role is the doping of helium clusters with NH_3 molecules to perform high-resolution spectroscopy. Such experiments are directed on the detection of non-classical response of molecular rotation in helium clusters addressing fundamental questions related to the microscopic nature of superfluidity. High-resolution spectroscopy on the NH_3–He complex is an important tool for increasing our understanding of intermolecular forces between NH_3 and He.

The experimental spectra of NH_3–He would be of great importance to test the accuracy of available potential energy surfaces used in the analysis of the role of resonances in low energy collisions [1, 2]. The rate coefficients of NH_3–He scattering are an important ingredient for numerical modeling of astrochemical environments. This is one of the reasons why NH_3–He interactions have been studied experimentally and theoretically by several groups (see [1, 2] and Refs. therein). Recently, a new four-dimensional potential energy surface (PES), based on a high-quality fit of 4180 *ab initio* points, was calculated [1]. The potential has a well depth of D_e = 35.08 cm^{-1}, which is to be compared with the well depth of 33.46 cm^{-1} for the earlier potential of Hodges and Wheatley [3]. Although this difference is not very large, it was found that small differences in the potential can have profound consequences for the observed resonance structures at low scattering energies. High-resolution spectra of the NH_3–He complex could be an additional test of the PES's quality, but these spectra have not been observed yet. High-resolution infrared and/or microwave spectra of the related NH_3–Ar (see [4] and Refs. there) and NH_3–Ne [5] in the gas phase have been reported, but no comparable data exist for helium.

A very different experiment in which the NH_3–He interaction plays an important role, is the trapping of NH_3 molecules inside He nanodroplets to perform high resolution spectroscopy. Such experiments are directed on the detection of a superfluid response of molecular rotation in helium clusters [6, 7]. It was shown that the lighter probe molecules allow one to measure the effective inertia of He clusters while maintaining a maximum

* Corresponding author: surin@isan.troitsk.ru

superfluid response with respect to dopant rotation [8]. Ammonia opens an additional possibility in such experiments to study the inversion motion in dependence of the helium cluster size. Infrared [9] and microwave [10] spectra of NH_3 in large (> 1000 He atoms) superfluid helium droplets were already observed.

Recently all bound rovibrational levels of the weakly bound NH_3–He complex have been calculated by A. van der Avoird [11] for total angular momentum $J = 0 \ldots 6$ on the new intermolecular potential surface [1]. The energy levels were obtained for two different nuclear spin states of NH_3–He arising from combinations of He with (*ortho*)-NH_3 and (*para*)-NH_3. The (*o*)-NH_3 and (*p*)-NH_3 complexes behave as different molecules, because they do not interconvert at the time scale of the experiment, due to the different nuclear spin statistics.

The rotation-inversion spectra of (*o*)-NH_3–He and (*p*)-NH_3–He in the frequency range of 5-25 GHz are under study guided by the bound state calculations. The experimental data on the rotational levels will be compared with the theoretical results.

The support from Russian Foundation for Basic Research (16-53-12061, 15-03-09333) and Deutsche Forschungsgemeinschaft (SCHN 1280/5-1) is acknowledged.

References

1. K.B. Gubbels, S.Y.T. van de Meerakker, G.C. Groenenboom, G. Meijer, A. van der Avoird, J. Chem. Phys. **136**, 074301 (2012)
2. O. Tkáč, A.J. Orr-Ewing, P.J. Dagdigian, M.H. Alexander, J. Onvlee, A. van der Avoird, J. Phys. Chem. **140**, 134308 (2014)
3. M.P. Hodges, R.J. Wheatley, J. Chem. Phys. **114**, 8836 (2001)
4. J. Loreau, J. Liévin, Y. Scribano, A. van der Avoird, J. Phys. Chem. **141**, 224303 (2014)
5. J. van Wijngaarden, W. Jäger, J. Chem. Phys. **115**, 6504 (2001)
6. J. Tang, Yu Xu, A.R.W. McKellar, W. Jäger, Science **297**, 2030 (2002)
7. L.A. Surin, A.V. Potapov, B.S. Dumesh, S. Schlemmer, Y. Xu, P.L. Raston, W. Jäger, Phys. Rev. Lett. **101**, 233401 (2008)
8. P.L. Raston, W. Jäger, H. Li, R.J. Le Roy, P.N. Roy, Phys. Rev. Lett. **108**, 253402 (2012)
9. M.N. Slipchenko, A.F. Vilesov, Chem. Phys. Lett. **412**, 176 (2005)
10. R. Lehnig, N.V. Blinov, W. Jäger, J. Chem. Phys. **127**, 241101 (2007)
11. A. van der Avoird, private communication

EPJ Web of Conferences **132**, 03046 (2017) DOI: 10.1051/epjconf/201713203046
SPECTROSCOPY.SU 2016

Millimeter-wave observations of the NH_3–D_2 and ND_3–H_2 van der Waals complexes

I. Tarabukin, V. Panfilov, A. Dolgov, and L. Surin*

Institute for Spectroscopy RAS, 108840 Troitsk, Moscow, Russia

Abstract. The rotational spectra of the ND_3–H_2 and NH_3–D_2 van der Waals complexes have been observed in a supersonic molecular jet for the first time. The OROTRON intracavity spectrometer was used to detect pure rotational transitions in the frequency range from 75 to 120 GHz. The search of the spectral lines was based on the initial observation of the main isotopic species of the NH_3–H_2. The measured transition frequencies including hyperfine splitting were analyzed in order to determine the energy levels, molecular parameters and structure of the *ortho*ND_3–*ortho*H_2 and *ortho*NH_3–*para*D_2 complexes.

Since its discovery in the interstellar medium in 1969, ammonia has been widely used as a probe of the physical conditions in a variety of interstellar environments, ranging from pre-stellar cores, molecular clouds, to external galaxies. The peculiar structure of the molecule makes ammonia lines excellent tracers of the density and temperature in these environments. Thus the excitation temperature between the first two *para*-metastable (rotational) levels is an excellent probe of the gas kinetic temperature. However, the calibration of this ammonia thermometer depends on the accuracy of the collisional rates with H_2. Although a number of collisional studies have been made on the NH_3-H_2 system, laboratory data generally do not directly provide state-to-state rate coefficients. As a result, radiative transfer models can exclusively rely on theoretical estimates. On the other hand, laboratory measurements are crucial to establish the predictive abilities of theory and, in particular, of the potential energy surfaces (PES). The NH_3-H_2 system is a van der Waals complex, and the bound states of such complexes are sensitive to the interaction potential [1]. As a result, high-resolution spectroscopy on the NH_3-H_2 complex is an important tool for increasing our understanding of intermolecular forces between NH_3 and H_2.

The study of the deuterated isotopologues provides an effective new probe of the intermolecular interaction between ammonia and dihydrogen and can help to test the existing *ab initio* potential energy surface calculations. In this work, the rotational spectra of the ND_3–H_2 and NH_3–D_2 van der Waals complexes have been observed in a supersonic jet for the first time. The millimeter-wave OROTRON intracavity jet spectrometer [1] along with double resonance technique was used to detect two nuclear spin species, *ortho*ND_3–*ortho*H_2 and *ortho*NH_3–*para*D_2, in the frequency range from 75 to 120 GHz. The spectra were predicted from the initial observation of the main isotopic species of the *ortho*NH_3–*ortho*H_2 [2]. The measured line positions including hyperfine splitting due to the ^{14}N nuclear spin of

* Corresponding author: surin@isan.troitsk.ru

ammonia were analyzed in order to determine the molecular parameters and structure of the *ortho*ND_3–*ortho*H_2 and *ortho*NH_3–*para*D_2 complexes.

The support from Russian Foundation for Basic Research (16-33-00616, 15-03-09333) is acknowledged.

References

1. P.E.S. Wormer, A. van der Avoird, Chem. Rev. **100**, 4109 (2000)
2. L.A. Surin, B.S. Dumesh, F. Lewen, D.A. Roth, V.P. Kostromin, F.S. Rusin, G. Winnewisser, I. Pak. Rev. Sci. Instrum. **72**, 2535 (2001)
3. L.A. Surin, I.V. Tarabukin, S. Schlemmer, A. Breier, T. Giesen, M.C. McCarthy, *Book of Abstracts of the 24th Colloquium on High Resolution Molecular Spectroscopy* (2015)

Relaxation of a three-level atom in stochastic fields and its spectroscopic properties

V.A. Mikhailov[1,*], and N.V. Troshkin[1]

[1]Samara National Research University, 443086 Samara, Russia

Abstract. Relaxation of a motionless three-level atom with a non-equidistant spectrum in a photon heat bath and an external stochastic field is investigated. In the Markovian approximation of the relaxation processes a master equation for the reduced density matrix is derived by making use of the averaging over realizations of the stochastic field. The master equation is solved exactly. The correlation functions, the shapes of the radiation lines, the intensity correlation and photon number spectrums are calculated for the V, Ξ and Λ atoms and realizations of the stochastic field by a delta-correlated and a Kubo-Anderson processes.

Spectroscopic properties of relaxation of two-level systems driven by external stochastic fields were studied in [1,2]. Here the case of a three-level system is investigated by means of a master equation averaged over stochastic fields realizations.

The Hamiltonian of a motionless three-level atom with a non-equidistant spectrum interacting with a photon heat bath and an external stochastic field has the next form [3]:

$$\hat{H} = \hat{H}_A + \hat{H}_T + \hat{H}_{AT} + \hat{H}_{st},$$

where $\hat{H}_A$ – the Hamiltonian of an isolated atom, $\hat{H}_T$ – the photon heat bath thermostat, $\hat{H}_{AT}$ – the Hamiltonian of the interaction between the atom and the heat bath.

The Hamiltonian of the interaction between the atom and the stochastic field can be written as:

$$\hat{H}_{st} = \hbar\left(\Omega(t)\hat{H}_1 + \Xi(t)\hat{H}_2\right) + \hbar\left(\xi(t)\hat{J}_+ + \zeta(t)\hat{K}_+ + \lambda(t)\hat{L}_+ + h.c.\right)$$

where $\hat{H}_1$ and $\hat{H}_2$ are diagonal operators defining energy levels of the atom; $\hat{J}_\pm$, $\hat{K}_\pm$ and $\hat{L}_\pm$ are operators of transitions between energy levels of the atom. The listed operators are generators of the *SU(3)* group. $\Omega(t)$ and $\Xi(t)$ define random shifts of atomic energy levels; $\xi(t)$, $\zeta(t)$ and $\lambda(t)$ are random functions proportional to intensity of the external stochastic field and affecting transitions between energy levels of the atom.

In the Markovian approximation of the relaxation processes a master equation for the reduced density matrix is derived by making use of the averaging over realizations of the stochastic field [4, 5, 6]:

* Corresponding author: va_mikhailov@mail.ru

$$\frac{\partial\langle\hat{\rho}\rangle}{\partial t} = K_\Omega\left(\hat{H}_1\langle\hat{\rho}\rangle\hat{H}_1 - \hat{H}_1^2\langle\hat{\rho}\rangle\right) + K_\Xi\left(\hat{H}_2\langle\hat{\rho}\rangle\hat{H}_2 - \hat{H}_2^2\langle\hat{\rho}\rangle\right) +$$
$$+(\eta_1 + K_\xi)\left(\hat{J}_-\langle\hat{\rho}\rangle\hat{J}_+ - \hat{J}_+\hat{J}_-\langle\hat{\rho}\rangle\right) + (\delta_1 + K_\xi)\left(\hat{J}_+\langle\hat{\rho}\rangle\hat{J}_- - \hat{J}_-\hat{J}_+\langle\hat{\rho}\rangle\right) +$$
$$+(\eta_2 + K_\zeta)\left(\hat{K}_-\langle\hat{\rho}\rangle\hat{K}_+ - \hat{K}_+\hat{K}_-\langle\hat{\rho}\rangle\right) + (\delta_2 + K_\zeta)\left(\hat{K}_+\langle\hat{\rho}\rangle\hat{K}_- - \hat{K}_-\hat{K}_+\langle\hat{\rho}\rangle\right) +$$
$$+(\eta_3 + K_\lambda)\left(\hat{L}_-\langle\hat{\rho}\rangle\hat{L}_+ - \hat{L}_+\hat{L}_-\langle\hat{\rho}\rangle\right) + (\delta_3 + K_\lambda)\left(\hat{L}_+\langle\hat{\rho}\rangle\hat{L}_- - \hat{L}_-\hat{L}_+\langle\hat{\rho}\rangle\right) + h.c.,$$

where

$$\eta_j = \frac{\gamma_j}{2}(N_j + 1), \quad \delta_j = \frac{\gamma_j}{2}N_j, \quad j = 1,2,3;$$

$$K_\Omega = \int_{t_0}^{t}\langle\Omega(t)\Omega(t_1)\rangle dt_1, \quad K_\Xi = \int_{t_0}^{t}\langle\Xi(t)\Xi(t_1)\rangle dt_1, \quad K_\xi = \int_{t_0}^{t}\langle\xi(t)\overline{\xi}(t_1)\rangle dt_1, \quad K_\zeta = \int_{t_0}^{t}\langle\zeta(t)\overline{\zeta}(t_1)\rangle dt_1,$$

$$K_\lambda = \int_{t_0}^{t}\langle\lambda(t)\overline{\lambda}(t_1)\rangle dt_1,$$

γ_j is the damping constant for the j-th transition of the atom, N_j is the average number of photons in the photon heat bath on the j-th transition of the atom.

The interaction of the atom with the external stochastic field is considered in the same order of smallness as the interaction of the atom with the photon heat bath.

Influence of model parameters of the stochastic field on shapes of radiation lines, spectrums of intensity correlations and number of photons is investigated for a delta-correlated process and a Kubo-Anderson process.

References

1. R. Walser, P. Zoller, Phys. Rev. A **49**, 5067 (1994)
2. I.A. Goychuk, Phys. Rev. E **51**, 6267 (1995)
3. V.A. Mikhailov, N.V. Troshkin, Proc. SPIE **9917**, 991731 (2016)
4. E.G. Petrov, Phys. Rev. E **57**, 94 (1998)
5. V.A. Mikhailov, Theoretical Physics **7**, 93 (2006)
6. V.A. Mikhailov, N.V. Troshkin, Computer Optics **40**, 61 (2016)

On large amplitude motions of simplest amides in the ground and excited electronic states

N.V. Tukachev*, V.A. Bataev, and I.A. Godunov

Lomonosov Moscow State University, 119991 Moscow, Russia

Abstract. For the formamide, acetamide, N-methylformamide and N-methylacetamide molecules in the ground (S_0) and lowest excited singlet (S_1) and triplet (T_1) electronic states equilibrium geometry parameters, harmonic vibrational frequencies, barriers to conformational transitions and conformer energy differences were estimated by means of MP2, CCSD(T), CASSCF, CASPT2 and MRCI ab initio methods. One-, two- and three-dimensional potential energy surface (PES) sections corresponding to different large amplitude motions (LAM) were calculated by means of MP2/aug-cc-pVTZ (S_0) and CASPT2/cc-pVTZ (S_1,T_1). For these molecules, in each excited electronic state six minima were found on 2D PES sections. Using PES sections, different anharmonic vibrational problems were solved and the frequencies of large amplitude vibrations were determined.

The simplest amides (RCONHR') in the ground electronic state prove to be suitable in modelling the structure and properties of peptides, which are abundant in nature and play crucial role in many biochemical processes. The investigation of the amides in the lowest excited electronic states is important considering different photophysical and photochemical processes in which amides take part. For instance, peptides and monosubstituted amides photoisomerize in aqueous solution to the less stable *cis* isomers during 220 nm ultraviolet resonance Raman experiments ([1]). Thus, the describing of the structure and conformational dynamics of the amide molecules in different electronic states is an important and actual problem.

The results presented in this paper are devoted to formamide ($HCONH_2$), acetamide (CH_3CONH_2), *N*-methylformamide ($HCONHCH_3$) and *N*-methylacetamide (CH_3CON-HCH_3) in the ground (S_0) and lowest excited singlet (S_1) and triplet (T_1) electronic states.

In S_0 electronic state the heavy atoms' frame of these molecules is planar though the potential functions of out-of-plane CNHR' vibration have a very shallow minima. Electronic excitation to S_1 and T_1 states was previously shown to cause strong pyramide-like deformation of both CNHR' and OCNR fragments and their rotation one around another in formamide and *N*-methylacetamide [2].

Among all amide vibrations, there are so-called large amplitude motions (LAM). These motions are known to play important role in describing the dynamics of conformationally non-rigid molecules. A few LAM could be considered in these molecules, namely: asymmetric (around "central" CN bond) and symmetric (the methyl top) internal rotations

* Corresponding author: tukachev.nikita@mail.ru

for all aforementioned electronic states, out-of-plane vibrations of CNHR' fragment in S_0 electronic state and the inversions of both CNHR' and OCNR fragments in excited electronic states.

In order to describe these LAM correctly, 1D, 2D and 3D vibrational problems were solved using following model Hamiltonian:

$$\hat{H}(\varphi_1,\dots,\varphi_n) = -\sum_{i=1}^{n}\sum_{j=1}^{n}\frac{\partial}{\partial\varphi_i}B_{ij}(\varphi_1,\dots,\varphi_n)\frac{\partial}{\partial\varphi_j} + V(\varphi_1,\dots,\varphi_n)$$

where n is the dimension of the problem (i.e. the number of vibrational degrees of freedom considered), φ_i — natural coordinates used in describing LAM, B_{ij} — kinematic parameters, and V — potential function (i.e. PES section).

Ab initio quantum chemistry methods used in this work for calculating various 1D, 2D and 3D PES sections along the LAM coordinates were MP2 (S_0) and CASPT2 (S_1, T_1) with basis sets aug-cc-pVTZ and cc-pVTZ, respectively. The active space (6-5) for the CASSCF and CASPT2 methods consisted of 5 molecular orbitals occupied by 6 electrons. These molecular orbitals are located on the carbonyl group and can be classified as σ_{CO}, π_{CO}, n_O, π^*_{CO}, σ^*_{CO}. All CASSCF and CASPT2 calculations were conducted as state specific.

Solving different vibrational problems with previously calculated PES sections as the potentials yields the vibrational levels' energies and vibrational wave functions. For S_0 electronic states anharmonic vibrational frequencies of formamide and acetamide are in good agreement with available experimental data [3,4]. For *N*-methylformamide, 2D (with internal rotation around the "central "CN bond and the NH out-of-plane vibration) and 3D vibrational frequencies of these two motions are in satisfactory agreement with each other and with harmonic approximation. However, the discrepancy revealed in the comparison of calculations and experiment [5] is significant enough to understand the importance of conducting more complicated and modern experiments [6]. For all considered molecules in S_0 electronic state there is a sufficient correlation between the internal rotation of "central" CN bond and out-of-plane CNHR' vibration.

On 2D PES sections of investigated amides in $^{1,3}(n,\pi^*)$ electronic states *ab initio* calculations predict six minima to exist (corresponding to 3 pairs of mirror symmetry conformers in the case of formamide and acetamide and to 6 entirely different conformers for *N*-methylformamide and *N*-methylacetamide). When compared to S_0 electronic state, in S_1 and T_1 electronic states the rotation around CN bond does not lead to essential changes in geometry structure of CNHR' fragment. In general, it could be said that the amide molecules in the lowest excited electronic states are less flexible than in ground electronic state. In excited electronic states of amide molecules increasing the dimension of vibrational problem solved (at least, from 1D up to 3D) does not lead to considerable changes of the lowest torsion levels energies. Vibrational wave functions in excited electronic states are more sophisticated than those in the ground electronic state: they are delocalized. This is the case due to close locations of PES minima and relatively low barriers to conformational transitions (compared to the ground electronic state).

The support by Russian Foundation for Basic Research is acknowledged (16-03-00794).

References

1. Y. Wang, J. Am. Chem. Soc. **111**, 8274 (1989)
2. Y. Li, R. L. Garrell, K. N. Houk, J. Am. Chem. Soc. **113**, 5895 (1991)
3. N.V. Tukachev et al., Comp. Theor. Chem. **1080**, 23 (2016)
4. N.V. Tukachev, V.A. Bataev, I.A. Godunov, J. Mol. Struct. (to be published)
5. S. Ataka, H. Takeuchi, M. Tasumi, J. Mol. Struct. **113**, 147 (1984)
6. N.V.Tukachev, V.A.Bataev, I.A.Godunov, J. Quant. Spec. Rad. Trans. **177**, 291 (2016)

DOI: 10.1051/epjconf/201713203049

Synthesis of $NaLuF_4:Er^{3+}$, Yb^{3+}, Ce^{3+} nanoparticles and study of photoluminescent properties in C - band

K.V. Khaydukov[1,*], V.V. Rocheva[1], A.G. Savelyev[1], M.E. Sarycheva[2], and I.M. Asharchuk[1]

[1]FSRC "Crystallography and Photonics" RAS, "Institute of Photonic Technology" RAS, 108840 Troitsk, Moscow, Russia
[2]Moscow Technological University, 119571 Moscow, Russia

Abstract. The novel core@shell nanocrystals β-$NaLuF_4$@$NaLuF_4$ co-doped with rare-earth ions Er^{3+}, Yb^{3+}, Ce^{3+} have been synthesized. The nano-particles indicate the intensive lines of anti-Stokes luminescence in the telecommunication C - band of spectrum when pumped at 970-980 nm. The nanoparticles have been characterized by transmission electron microscopy and spectrofluorimetry. The nanoparticles have a size 40-80 nm and possess the intensive photo-luminescence 73 nm bandwidth centered around 1530 nm. The photo-luminescence kinetics of β-$NaLuF_4$: Er^{3+}/ Yb^{3+}/ Ce^{3+} has been studied in IR range of spectrum. We have demonstrated that doping with cerium ions prevents serial stepwise excitation of erbium ions. Consequently, the lifetime of transition in erbium $^4I_{13/2}\rightarrow{}^4I_{15/2}$ has risen up to 6.9 ms. Intensity of 1530 nm line in Er^{3+} ions excited at 980 nm has been increased up to 6 times. Therefore, the nanoparticles are applicable to fabrication of compact waveguide amplifiers for C - band.

Nano-sized phosphor crystals form a new generation of nanomaterials as they possess the unique optical properties attractive for applications in telecommunications technology, integrated optics and photonics. The unique luminescent properties of nanophosphors are the result of successive absorption of photons by ytterbium ions and resonant non-radiative energy transfer to the erbium ions within the matrix of the hexagonal β-$NaLuF_4$ phase inorganic nanocrystal [1]. Figure 2 shows the energy level diagram of the crystal $NaLuF_4$: Er^{3+}/ Yb^{3+}/ Ce^{3+} explaining the down-conversion mechanism of 980 nm radiation. The pump photon at 980 nm is absorbed by the ion Yb^{3+}, transferring the latter into the excited state $^2F_{5/2}$. As a result of non-radiative resonance energy transfer, level $^4I_{13/2}$ of Er^{3+} is populated with the possibility of subsequent excitation of overlying energy states. Ions Ce^{3+} are capable to resettle resonant energy states from Er^{3+} ions above the state $^4I_{13/2}$. Therefore, the cerium ions act as constraints impeding the up-conversion process in the nanoparticle. Thus, the maximum efficiency of nanoparticle excitation is achieved with the following radiative decay at 1530 nm wavelength [2].

* Corresponding author: haidukov_11@mail.ru

Hexagonal shape nanocrystals have a size distribution in the range 40 - 80 nm (see the inset in Fig. 1). The photoluminescence of nanoparticles has been studied using spectrofluorimeter Fluorolog 3 (HJY, France) under the excitation by semiconductor laser at 977 nm. The luminescence spectra, measured at room temperature, have an intense line in the NIR (see Fig. 1). The utilization of more effective β-$NaLuF_4$ matrix in comparison to β-$NaYF_4$ makes possible to enhance the quantum yield from 2 to 6%. The introduction of cerium ions into the nanocrystal matrix allows to increase the efficiency of transition $^4I_{13/2} \rightarrow {}^4I_{15/2}$ up to 6 times. The transition lifetime is 6.9 ms (see Fig. 3).

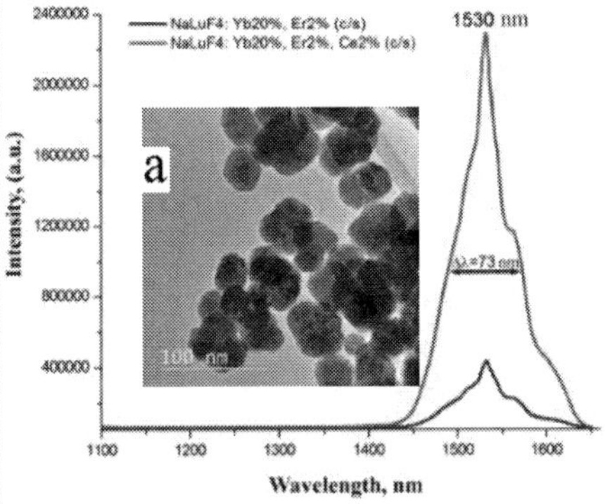

Fig. 1. TEM image (the inset) and the PL spectra of synthesized β-$NaLuF_4$: Er^{3+}, Yb^{3+}, Ce^{3+} and β-$NaLuF_4$: Er^{3+}, Yb^{3+} nanocrystals under excitation at 977 nm.

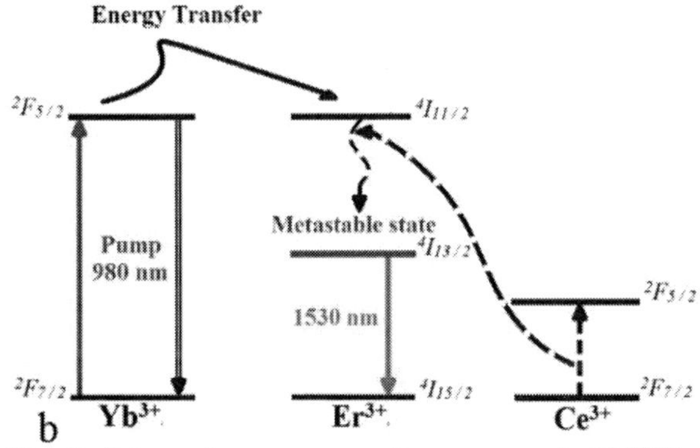

Fig. 2. Schematic energy transfer processes in NaYF4 crystal between Er^{3+}, Yb^{3+} and Ce^{3+}.

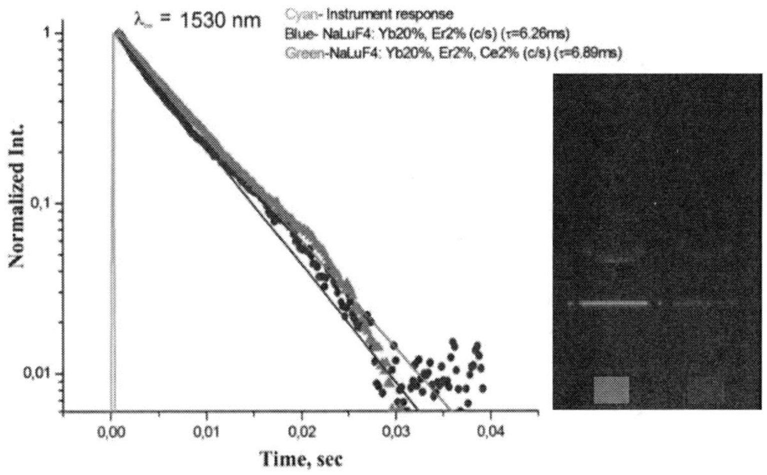

Fig. 3. Kinetics of photoluminescence decay of β-$NaLuF_4$: Er^{3+}, Yb^{3+}, Ce^{3+} and β-$NaLuF_4$: Er^{3+}, Yb^{3+} around 1.5 μm under the excitation at 977 nm. Excitation pulse length duration is 80 μs. The inset demonstrates two cuvettes with nanoparticles dissolved in dichloromethane (CH_2Cl_2) and illuminated by 975 nm laser beam.

Results of the conducted research indicate that the synthesized nanoparticles are a promising platform for fabrication of compact waveguide amplifiers for C - band.

The work was supported by Grant RFBR # 16-32-00935-mol_a.

References

1. A.N. Generalova et al., Nanoscale **7**, 1709 (2015)
2. X. Zhai et al., J. of Materials Chemistry C **3**, 1555 (2013)

EPJ Web of Conferences **132**, 03050 (2017) DOI: 10.1051/epjconf/201713203050
SPECTROSCOPY.SU 2016

A compact two-wave dichrometer of an optical biosensor analytical system for medicine

D.P. Chulkov[1], V.M. Gusev[1], O.N. Kompanets[1,*], F.V. Vereschagin[1], S.G. Skuridin[2], and Yu.M. Yevdokimov[2]

[1]Institute for Spectroscopy RAS, 108840 Troitsk, Moscow, Russia
[2]Engelgardt Institute of Molecular Biology RAS, 119991 Moscow, Russia

Abstract: An experimental model has been developed of a compact two-wave dichrometer on the base of LEDs that is well-suited to work with "liquid" DNA nanoconstructions as biosensing units. The mobile and inexpensive device is intended for use in a biosensor analytical system for rapid determination of biologically active compounds in liquids to solve practical problems of clinic medicine and pharmacology.

We develop optical biosensor analytical systems, in which "liquid" particles of the cholesteric liquid-crystalline dispersion (CLCD) of DNA are used as a biosensing unit, and a portable dichrometer registers changes of the circular dichroism (CD) signal generated with the DNA biosensing unit during its interaction with a defined biologically active compound (BAC) [1]. DNA biosensing units exhibit anomalously large optical activity in the CD spectrum in an absorption band of DNA nitrogen bases (chromophores) at 270 nm, the magnitude of which is stable and acts as a convenient "quality" criterion of prepared DNA CLCD particles. Intercalation of coloured BACs into the DNA cholesteric liquid-crystalline structure leads to the appearance in the CD spectrum of an additional abnormal band in the visible spectrum, whose amplitude depends on the concentration of BAC - this calibration dependence is used then to establish the presence of such compounds in the assay medium and to determine their concentration.

Among our developments there are dichrometers with tunable wavelength light sources [2, 3] which, due to their versatility, are primarily used for research purposes. Taking into account the specificity of the CD spectrum generated by biosensing units based on "liquid" DNA CLCD particles at intercalation therein of coloured BAC, the idea has occured to develop a compact biosensor analytical system for the rapid detection in clinical conditions of several important BAC, in particular, antibiotics (daunorubicin and analogues) used in oncological practice. Such the biosensor would work on only two discrete wavelengths, one of which corresponds to the absorption band of DNA chromophores at a wavelength of 270 nm, and the other – to the abnormal band in the visible region of the CD spectrum of the (DNA + BAC) complex. Such a choice predetermined steps to simplification of the dichrometer - a rejection of expensive broadband lamp (Xe) of the radiation source, a monochromator and a photomultiplier. The use of modern high power light-emitting diodes as light sources and a photodiode for registering the CD signal allows both DNA CLCD particles quality control at a wavelength of 270 nm and registration of the smallest changes

* Corresponding author: onkomp@isan.troitsk.ru

of the abnormal CD signal in the visible region resulting of interaction of the detectable BAC molecules with DNA CLCD particles with formation of the (DNA + BAC) complex.

The experimental two-wave dichrometer model that is intended to determine the daunomycin (DAU), for example, uses high brightness light-emitting diodes UVTOP270TO39BL (Sensor Electronic Technology Company) operating at 270 nm, and two LEDs (Prolight Company): the main one - S12LG2C-B emitting green light at 510 nm corresponding to the abnormal band in the CD spectrum of the (DNA + DAU) complex, and the auxiliary one - S12LR2C-B emitting red light at 634 nm and serving for fixing the "CD basic line". Radiation of LEDs is aligned along the same path by dichroic mirrors set at an angle of 45° to the directions of the emitters light flows, all sources are turned on alternately. To compensate the divergence of the radiation collimating lenses are used, and for narrowing the emission spectrum to 3-6 nm interference filters are applied.

In addition to the LED light sources and optics to them, the two-wave dichrometer model uses the standard for the previous multifunction dichrometers [2] elements: a linear polarizer, a photoelastic modulator of circular polarization of radiation, and a temperature controlled (0.5 °C accuracy within the temperature range of 20 - 80 °C) cuvette compartment for placement in a quartz spectrophotometric cuvette (10x10 mm) of test samples - "liquid" biosensing unit based on DNA CLCD particles in contact with the test liquid containing the detectable BAC. To register the anomalous CD signal and its changes the photodiode S1337-1010BQ (Hamamatsu Company) is applied.

Fig. Photo of the compact experimental two-wave dichrometer model adapted to work as a part of the DNA based biosensor analytical system

The two-wave dichrometer model is implemented on the basis of a built-in computer. The motherboard VIA EPIA PX ("VIA Technologies" Company) of form factor Pico-ITX which is a full-featured motherboard for x86 processors combines multiple units of the control system. A seven-inch widescreen color LCD VGA monitor with touch screen is used to display graphic information and provide the user interface. The device has broad communication capabilities: depending on the user requirements it can transfer data over wireless channels, Wi-Fi and Bluetooth, as well as an Ethernet network.

The two-wave dichrometer model dimensions have been reduced to 210x314x350 mm, its full power consumption does not exceed 100 Watts. The developed two-wave dichrometer model, being relatively compact, without any loss of sensitivity of the CD signal registration can be adapted to work in the biosensor test system in five modes: calibration of the system in the UV and visible regions of the spectrum; testing of optical properties of BAC; calibration of the optical properties of biosensing units based on DNA CLCD particles, and control of their optical activity prior to use; study of the dynamics of DNA nanoconstructions transformation during their interaction with BAC; determining the presence and concentration of BAC in the test fluid.

In the last regime mode the dichrometer operates by means of the program "Analytics" which, through a dialogue, "leads" the user through all steps prescribed with the laboratory technique of measuring a BAC concentration in the test liquid by circular dichroism (developed by the EIMB), and outputs the final result value as the BAC concentration.

The support from the Program "Basic researches for the developing biomedical technologies" of the Presidium of RAS is acknowledged.

References

1. Yu.M. Yevdokimov, V.I. Salyanov, S.G. Skuridin, *Nanostructures and nanoconstructions based on DNA* (CRC Press, New York, 2012)
2. O.N. Kompanets, Yu.M. Yevdokimov, Physics Uspekhi **52**, 304 (2009)
3. Yu.M. Yevdokimov, O.N. Kompanets, Physics Uspekhi **57**, 615 (2014)

EPJ Web of Conferences **132**, 03051 (2017) DOI: 10.1051/epjconf/201713203051
SPECTROSCOPY.SU 2016

Emission spectra of YAG:Er^{3+} under pulse laser-thermal excitation

V.M. Marchenko, and Yu.A. Shakir*

Prokhorov General Physics Institute of Russian Academy of Sciences, 119991 Moscow, Russia

Abstract. Spectra and kinetics of emission of YAG:0.5% Er^{3+} monocrystal in visible and NIR ranges were investigated under laser-thermal excitation by the pulses of CO_2 laser of 100 ns duration at wavelength $\lambda = 10{,}6$ μm. Kinetics of integral emission was interpreted.

Spectra of YAG:0.5% Er^{3+} monocrystal under laser-thermal excitation by pulses of CO_2 laser were recorded by spectrometer AvaSpec 2048 with sensitivity calibrated using etalon tungsten lamp. Recording time of monocrystal emission was 1 s. Integral emission of monocrystal was transmitted by optical fiber on photodiode LFD-2A with sensitivity region of 400-1700 nm and recorded by oscilloscope Tektronix TDS 220.

Laser-thermal spectra of monocrystal are superpositions of continuous spectrum and selective emission (SE) peaks. Intensities of spectra grow with increase of laser pulse energy densities $W_1<W_2<W_3$ (Fig.1). Positions of SE peaks (1, 2, 3) in region of λ=544÷999 nm correspond to Er^{3+} transitions $^4I_{9/2}$, $^4I_{11/2} \rightarrow {}^4I_{15/2}$ observed in luminescence spectrum of the monocrystal (Fig.2) also. Separate spectral peaks (2, 3) in region of λ=397÷660 nm on Fig.1 correspond to wavelengths of plasma radiation peaks of air laser-induced breakdown spectra [3, 4].

For interpretation of kinetics of YAG:Er^{3+} monocrystal emission we used model of unsteady transport of energy by both heat conduction and self-emission in plane layer with the monocrystal parameters [1, 2]. According to calculations input layer of 100 μm thickness is heated by the laser pulse with W=0.5 ÷ 2.0 J/cm^2 and λ=10.6 μm up to T= 500 ÷ 1100K and population of Er^{3+} electronic-vibrational levels increases. Calculated space-time distribution of temperature T(z, t) was used in kinetics equations of population of $^4I_{15/2}$ и $^4I_{13/2}$ levels in the input layer (z=0). Temporal dependence $T(0, t)/T_0$ (T_0=293K) is presented on Fig. 3 (2). Solution of kinetics equations gave SE intensity I(t) of the selected transition of Er^{3+} (presented on Fig. 3 (3)) which corresponds to oscillogram of monocrystal integral emission (4) on Fig. 3. This fact is result of thermal equilibrium population of Er^{3+} levels in the layer.

* Corresponding author: shakir@kapella.gpi.ru

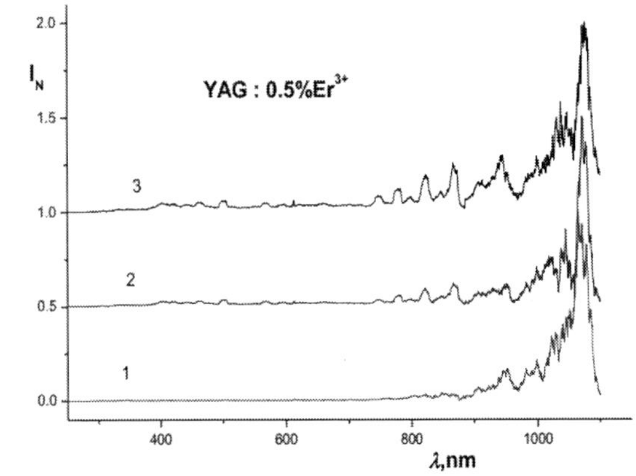

Fig. 1. Normalized emission spectra (1, 2, 3) of monocrystal YAG: 0.5%Er^{3+} under $W_1<W_2<W_3$ energy densities of pulse laser-thermal excitation at λ_{ex} = 10.6 μm.

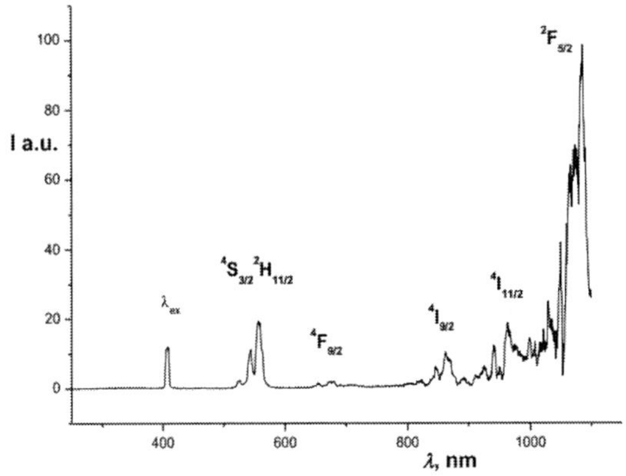

Fig. 2. Luminescence spectrum of YAG: 0.5%Er^{3+} monocrystal under laser diode excitation at λ_{ex} = 405 nm.

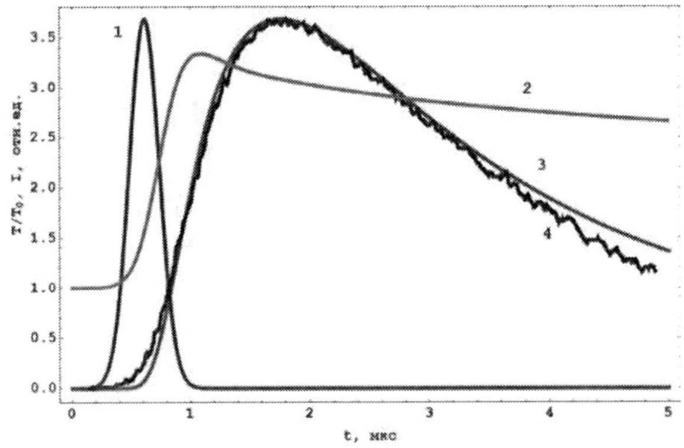

Fig. 3. Calculated temporal dependences of temperature T/T_0 (2) and SE intensity (3) of YAG: Er^{3+} monocrystal under laser pulse (1) excitation. Oscillogram of integral emission (4).

References

1. V.M. Marchenko, Y.A. Shakir, *International Conference "Laser Optics – 2014"* (St. Petersburg, 2014)
2. V.M. Marchenko, Y.A. Shakir. Advances in Applied Physics **2**, 335 (2014)
3. R.J. Nordstrom, Appl. Spectroscopy **49**, 1490 (1995)
4. L.J. Radziemski, D.A. Cremers, M. Bostian, R.C. Chinni, C. Navarro-Northrup, Appl. Spectroscopy **61**, 1141 (2007)

Influence of transannular interaction over absorption and fluorescent properties of [2.2] paracyclophane and its phenyl derivatives

R.N. Nurmukhametov[1], A.V. Shapovalov[2,*], and D.Yu. Antonov[2]

[1]L.Ya. Karpov Institute of Physical Chemistry RAS, 105064 Moscow, Russia
[2]A.N. Nesmeyanov Institute of Organoelement Compounds RAS, 119991 Moscow, Russia

Abstract. A significant bathochromic shift of the fluorescent and long-wavelength absorption bands of [2.2] paracyclophane comparing to corresponding bands of alkyl-benzenes is due to a strong transannular interaction, resulting in formation of a principally new excited state of lower energy. It is concluded that the fluorescent levels for alkylbenzene excimers and for the macrocycle are of the same nature. Analysis of [2.2] paracyclophane mono- and diphenylderivatives spectra shows that their intensive absorption bands (230–310 nm) are originated from electron transitions of biphenyl groups and weak long wavelength absorption (310–340 nm) and fluorescent bands are governed by the same electron transitions between ground and excimer-like excited states as in the case of non-substituted macrocycle.

It is well known, that UV-irradiation of organic luminophores solutions (included solid solutions in polymer) initiates in these systems photophysical and photochemical processes, which are essentially depended on interactions between molecules in excited and non-excited states. In one cases this leads to excited dimmers (excimers) formation, in the others – to concentration decay of fluorescence or photochemical reactions. For organic compounds these processes cause non-radiative degradation of singlet excited state, which leads to fluorescence efficiency decrease in comparing to the single molecules emission. To investigate how these factors influence on the fluorescence of organic molecules, systems that contain several fragments with π-electron system in the certain orientation to each other are commonly used. In case when two chromophoric groups are included into cycle and located at a low distance non-valent interaction (so called "transannular interaction") originates. Due to this interaction absorption and luminescent properties of the macrocycles are considerably different in comparison with the properties of their non-cycled analogues. [2.2] Phanes are compounds possessed two chromophoric groups at a distance about 3Å in "face-to-face" position by means of ethylene linkages. Such macrocycles let us to investigate dimmer or excimer emission of sandwich-like compounds and polymers.

It is estimated that the main reason of significant bathochtomic shift of [2.2] paracyclophane ([2.2] PCP) fluorescent and long-wavelength absorption bands compared with the corresponding bands for molecules containing non-conjugated benzene chromophores is a strong transannular interaction between π-orbitals of closely spaced benzene rings of the macrocycle. This interaction leads to formation of principally new

* Corresponding author: Shapovalov-alex@yandex.ru

 DOI: 10.1051/epjconf/201713203052

excited state of lower energy. It is concluded that the fluorescent state for the macrocycle and for alkyl-benzenes excimers are of the same nature. It is supported by the fact that [2.2] PCP characterizes by very low quantum yields (~1 %) and fluorescence lifetimes (~10 ns) as well as excimer emission of benzene and its methyl derivatives (corresponding parameters for alkyl-benzenes molecular fluorescence are markedly greater: ~10 % and ~25 ns). Another characteristic feature of [2.2] PCP and alkyl-benzenes excimer fluorescence is their featureless emission bands even at low temperatures. The [2.2] PCP fluorescence band location at longer wavelength compared with polystyrene excimer emission band is explained by the fact that macrocycle HOMO energy level is situated higher than that of alkyl-substituted benzenes (the difference between [2.2] PCP and *p*-xylene ionization potentials is equal to 0.34 eV, 2720 cm^{-1}, which agrees well with the mentioned bathochromic shift – 3130 cm^{-1}) [1]. That is why no anomalously large Stokes shift (one of the excimer emission characteristic signs) for [2.2]PCP fluorescence band can be observed.

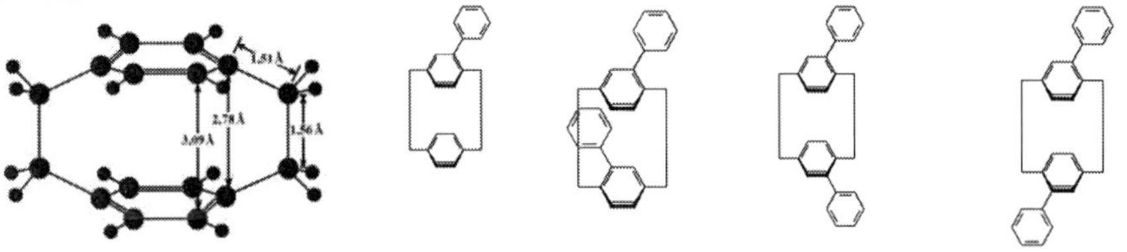

a) I II III IV

Fig. 1. [2.2] PCP molecular structure (a) and its derivatives (I – IV).

[2.2] PCP phenyl derivatives **I** – **IV** are interesting objects for electron spectroscopy because the paracyclophane chromophore is retained in them and they also contain other groups with π-conjugated system. Analysis of **I** – **IV** absorption spectra shows that they are determined by two types of chromophores: biphenyl and paracyclophane. It is shown that their weak long wavelength absorption bands at 310–340 nm are governed by paracyclophane chromophore and strong bands at 230–310 nm originate from electronic transitions of biphenyl groups. It is supposed that observed difference of isomeric macrocycles **II** – **IV** strong bands is caused by different resonance interaction between electron oscillators (transitions) of the two biphenyl chromophores leading to Davydov splitting of their excited states [2]. The magnitude of the splitting is inversely proportional to the distance between the molecules. This distance in single crystal of benzene is slightly greater than that between the chromophores in **II** – **IV**, therefore resonance interaction in macrocycles is expected to be more pronounced. It is demonstrated that electronic transition energy of the paracyclophane chromophore changes little upon phenyl substitution: fluorescence spectra of **I** – **IV** solutions at 298 and 77K exhibit featureless bands with shapes similar to the band of [2.2] PCP solution, slightly shifted to it bathochromically for 10 – 20 nm (like their long wavelength absorption bands) [3]. **I** – **IV** quantum yields are 2 – 3 times higher than that of the non-substituted compound and their fluorescence lifetimes are approximately the same. It also supports that their $S_1 \leftrightarrow S_0$-transition is determined by paracyclophane chromophore. It can be concluded that long-wavelength absorption and fluorescent properties of [2.2] PCP phenyl derivatives are determined by excimer-like excited state that essentially forms due to transannular interaction (as for non-substituted [2.2] PCP). These results will facilitate interpretation of other cyclophanes π-electron features which are essential for understanding the nature of photophysical and photochemical processes in polychromophoric systems.

References

1. R.N. Nurmukhametov, A.V. Shapovalov, A.M. Sergeev, J. Appl. Spectr. **81**, 49 (2014)
2. A.S. Davydov, *Theory of Molecular Excitons* (Nauka, Moscow, 1968) (in Russian)
3. R.N. Nurmukhametov et al., J. Appl. Spectr. **83**, 27 (2016)

EPJ Web of Conferences **132**, 03053 (2017) DOI: 10.1051/epjconf/201713203053
SPECTROSCOPY.SU 2016

An XPS method for layer profiling of NbN thin films

A.V. Lubenchenko[1,*], A.A. Batrakov[1], A.B. Pavolotsky[2], S. Krause[2], I.V. Shurkaeva[1], O.I. Lubenchenko[1], and D.A. Ivanov[1]

[1]Moscow Power Engineering Institute, 111250 Moscow, Russia
[2]Chalmers University of Technology, 41296 Göteborg, Sweden

Abstract. Layer chemical and phase profiling of niobium nitride thin films on a silicon substrate oxidized on air was performed with the help of a method designed by us. The method includes: a new method of background subtraction of multiple inelastically scattered photoelectrons considering depth inhomogeneity of electron inelastic scattering; a new method of photoelectron line decomposition into component peaks considering physical nature of different decomposition parameters; joint solution of the background subtraction and photoelectron line decomposition problems; control of line decomposition accuracy with the help of a suggested performance criterion; calculation of layer thicknesses for a multilayer target using a simple formula.

The interest in thin films of niobium and its compounds is caused by their superconductive properties used in cryogenic electronic instrumentation. For example, niobium nitride thin films are attracting rising attention in recent submillimeter and THz receiver research [1]. As an oxidation result of films of niobium or its compounds (originally homogeneous) multilayer and multiphase films are built. The superconductive properties of niobium compounds become worse with increase of surface niobium oxide layer thickness. Oxide films of niobium and its compounds are covered by a great amount of works, i.e. [2-5]. However, there are no works covering layer phase profiling of oxidized thin films of niobium compounds.

This work presents layer phase profiling of oxidized thin films of niobium nitride on a silicon substrate by the means of XPS. Thin niobium nitride films were sputtered onto a silicon substrate by the magnetron method. The films were oxidized during unload from the vacuum chamber. X-ray photoemission spectra were obtained with the electron-ion spectroscopy module based on the platform Nanofab 25 (NT-MDT). In the analytical chamber, an ultrahigh oil-free vacuum about 10^{-8} Pa was achieved. The spectra were recorded with an electrostatic semi-spherical energy analyzer SPECS Phoibos 225.

Layer profiling on the base of XPS spectra of multilayer multicomponent thin films interpretation is a complex inverse problem with a lot of unknown parameters. For correct solution of this problem new methods and approaches designed by the authors are suggested:

* Corresponding author: LubenchenkoAV@mpei.ru

- the background subtraction method considering the difference between surface and volume energy losses;
- using of parameters that are constant in all the range of XPS for calculation of background and line profiles;
- using of parameters of line profiles defined from spectra of the Handbook of X-Ray Photoelectron Spectroscopy for pure homogeneous targets;
- calculation formulae for natural peak widths of photoelectrons from chemically bound elements;
- joint interpretation of different lines of one element using the same model;
- calculation of layer thicknesses and chemical and phase compound in the multilayer target model considering the results of decomposition of various lines.

On the base of the designed methods, layer chemical and phase profiling of air oxidized niobium an niobium nitride films on a silicon substrate was performed. The results of this analysis:

- a *niobium film 100 nm thick*: on the film surface there is a hydrocarbon layer 1.4 nm thick, oxide Nb_2O_5 layer 9.6 nm thick, under this layer there is a suboxide Nb_2O layer 1 nm thick, then there is metallic niobium.
- a *NbN film 10 nm thick*: on the film surface there is a hydrocarbon layer about 1 nm thick. Then there is a multiphase oxide layer of NbO_x (70% Nb_2O_5, 7% Nb_2O_3, 23% NbO) with a total thickness of 2 nm. This layer is generated after unloading the target from the magnetron chamber into air. During the NbN thin film oxidation process the phase state of niobium nitride changes. Under the oxide layer a niobium nitride layer of another phase NbN_x (probably Nb_5N_6) 1.5 nm thick is generated. Then there is an unchanged layer of NbN.

References

1. D. Buchel et al., IEEE Transact. on Terahertz Science and Technology **5**, 207 (2015)
2. A. Darlinski, J. Halbritter, Surface and Interface Analysis **10**, 223 (1987)
3. Z. Weibin et al., Surface and Interface Analysis **45**, 1206 (2013)
4. H. Liu, N. Gao, M. Liao, X. Fang, Scientific reports **5**, 7716 (2015)
5. S. Krause et al., IEEE Transact. on Applied Superconductivity **26**, 1 (2016)

Spectroscopy of photonic band gaps in mesoporous one-dimensional photonic crystals based on aluminum oxide

V.S. Gorelik[1,*], Yu.P. Voinov[1], V.V. Shchavlev[2], Dongxue Bi[3], Guo Liang Shang[4], and Guang Tao Fei[4]

[1]Lebedev Physical Institute RAS, 119991 Moscow, Russia
[2]Moscow Institute of Physics and Technology, 141700 Dolgoprudny, Russia
[3] Bauman Moscow State Technical University, 105005 Moscow, Russia
[4] Hefei Institutes of Physical Science, Chinese Academy of Sciences, 230031 Hefei, China

Abstract. Mesoporous one-dimensional photonic crystals based on aluminum oxide have been synthesized by electrochemical etching method. Reflection spectra of the obtained mesoporous samples in a wide spectral range that covers several band gaps are presented. Microscopic parameters of photonic crystals are calculated and corresponding reflection spectra for the first six band gaps are presented.

At present, photonic crystals [1] are intensively investigated. A characteristic feature of such materials is the presence of photonic band gaps (stop bands) in their spectrum. The simplest example of such structures is a one-dimensional photonic crystal composed of the layers of alternating thickness a_1, a_2 and refractive index n_1, n_2 [2]. One of the methods to manufacture such structures is the technique of electrochemical etching [3]. Mesoporous photonic crystal films based on anodic Al_2O_3 are interesting for exploration [4-6]. These structures demonstrate high melting temperature, chemical inertness, high thermal conductivity, etc. In this study, the samples of one-dimensional aluminum-oxide-based photonic crystal films were synthesized with different crystal lattice periods in the range of 350-450 nm. These samples provide photonic stop bands in a wide spectral range from the near IR to UV.

Fig. 1. (a) Surface of the photonic crystal film obtained with frontal shooting configuration; a scanning electron microscopy image of the of the film upper surface (b) and its lateral side surface (c). Scanning scale division is 0.1 μm.

A photograph of one of the samples obtained for frontal shooting configuration is shown at Fig. 1a. Photonic-crystal aluminum oxide film changes its color (iridescent) when we change the viewing angle. Fig. 1 (b, c) shows the microscopic structure of the front and lateral side surfaces of the film. The analysis of these surfaces provided the values of the pore

* Corresponding author: gorelik@sci.lebedev.ru

diameter d = 55-85 nm, the porosity δ = 0.45 of the upper layer of the sample (Fig. 1b) and the period a = 386 nm of the photonic crystal (Fig. 1c). The spectra of broadband optical radiation reflected from the surface of photonic crystal on anodic Al_2O_3 were recorded. The reflectance spectra corresponding to the first three photonic bang gaps were obtained by different manner; the results are shown in Fig. 2a.

The theory [7] was used to describe optical properties of the periodic structures and establish corresponding dispersion relation for electromagnetic waves in one-dimensional periodic structures. The dependence of reflection coefficient on wavelength was calculated with the use of the obtained dispersion relation with the dispersion of the refractive indices of the layers taken into account (Fig. 2b). Such spectrum covers first six photonic stop bands of the investigated photonic crystal.

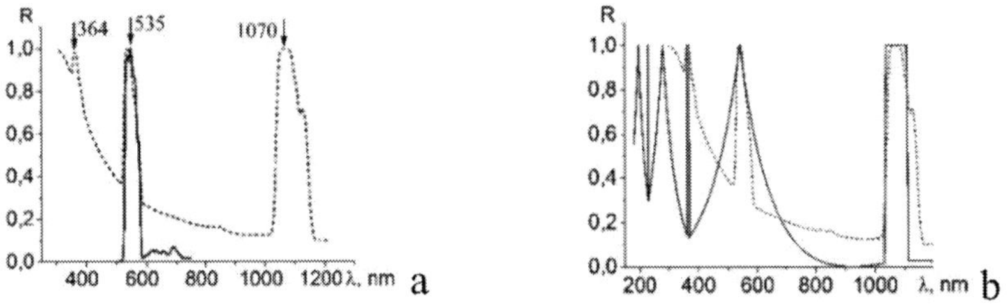

Fig. 2. (a) The reflection spectrum for a mesoporous one-dimensional Al_2O_3 photonic crystal obtained experimentally; (b) The reflection spectra obtained experimentally (dotted line) and the calculation results (solid line).

Table 1 provides a comparison between the calculation results and experimental data for the considered photonic crystal.

Table 1. Characteristics of the photonic crystal film sample obtained experimentally (exp) and theoretically (th) (m - number of stop band; λ_{exp}, λ_{th} - the spectral position of the center of the stop band; $\Delta\lambda_{exp}$, $\Delta\lambda_{th}$ - the width of the stop band; $n_{ef}(\lambda_m)$, $n_1(\lambda_m)$, $n_2(\lambda_m)$ - the effective refractive index and the refractive indices of the layers; a_{exp}, a_{th} - the period of the structure)

m	λ_{exp}, nm	λ_{th}, nm	$\Delta\lambda_{exp}$, nm	$\Delta\lambda_{th}$, nm	$n_1(\lambda_m)$	$n_2(\lambda_m)$	$n_{ef}(\lambda_m)$	a_{exp}, nm	a_{th}, nm
1	1072	1070	78	78	1.464	1.305	1.387	386	385.8
2	535	539	41	40	1.475	1.314	1.397	386	385.8
3	364	365	–	24	1.491	1.344	1.415	386	385.8
4	–	279	–	16	1.510	1.380	1.446	386	385.8
5	–	228	–	13	1.541	1.411	1.477	386	385.8
6	–	196	–	9	1.579	1.467	1.524	386	385.8

Thus, the proposed scheme of determining the values of the microscopic characteristics of the mesoporous photonic crystal allows to obtain the reflection spectra and use them for determination of the spectral positions of the photonic stop bands. The synthesized photonic crystal films can be used as notch filters in the infrared and visible regions of the spectrum, as selective laser resonator mirrors and substrates for the effective sensors of molecular structures.

References

1. E. Yablonovich, Phys. Rev. Lett. **58**, 2059 (1987)
2. V.S. Gorelik, Quant. Electron. **37**, 409 (2007)
3. L. Pavesi, Rivista del Nuovo Cimento **20**, 1 (1997)
4. L.Yisen, Y. Chang, L. Zhiyuan, L. Yi, H. Xing, Electrochem. Commun. **13**, 1336 (2011)
5. V.S. Gorelik, V.V. Shchavlev, Bull. of the Lebedev Physics Institute **43**, 189 (2016)
6. V.S. Gorelik et al., Opt. Spectrosc. **120**, 534 (2016)
7. A. Yariv, P. Yeh, *Optical waves in crystals: Propagation and control of laser radiation* (Wiley, New York, 1984)

Hadronic deuteron polarizability contribution to the hyperfine structure in muonic deuterium

A.V. Eskin[1,*], A.P. Martynenko[1], and E.N. Elekina[2]

[1]Samara University, 443086 Samara, Russia
[2]Samara State University of Architecture and Civil Engineering, 443001 Samara, Russia

Abstract. The calculation of the contribution to the polarizability of the nucleus to hyperfine structure of muonic hydrogen is carried out within the unitary isobar model and on the basis of experimental data on the structure functions of deep inelastic lepton-proton and lepton-deuteron scattering. The calculation of virtual absorption cross sections of transversely and longitudinally polarized photons by nucleons in the resonance region is performed in the framework of the program MAID.

Hadronic deuteron-polarizability contribution of the order $(Z\alpha)^5$ to the Lamb shift and hyperfine structure is determine of the virtual forward Compton scattering $\gamma^* + d \to \gamma^* + d$ (see Fig.1). This amplitude contain deuteron tensor [1-3]:

$$M_{\mu\nu}^{(d)} = \left(-g_{\mu\nu} + \frac{k_\mu k_\nu}{k^2}\right) C_1(\nu, k^2) + \frac{1}{m_d^2}\left(p_{2\mu} - \frac{m_d \nu}{k^2} k_\mu\right)\left(p_{2\nu} - \frac{m_d \nu}{k^2} k_\nu\right) C_2(\nu, k^2) + i\varepsilon_{\mu\nu\alpha\beta} k^\alpha \left\{m_d S^\beta H_1(\nu, k^2) + \left[(P \cdot k) S^\beta - (S \cdot k) P^\beta\right] \frac{H_2(\nu, k^2)}{m_d}\right\}, \quad (1)$$

where k is the virtual photon 4-momentum, $\nu = k_0$ is the virtual-photon energy. The symmetric part of the tensor in (1) contributes to the Lamb shift [4-5] (structure functions $C_{1,2}(\nu, k^2)$), while its antisymmetric part contributes to the hyperfine structure (structure functions $H_{1,2}(\nu, k^2)$) [2-3].

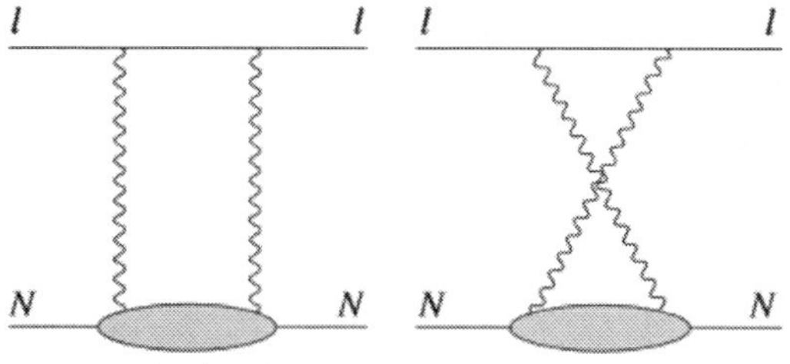

Fig. 1. Two-photon Feynman amplitudes determining the correction on nucleus polarizability

The polarized structure functions of proton and neutron $G_1^{p,n}(\nu, Q^2)$ and $G_2^{p,n}(\nu, Q^2)$ determine a correction of the hadronic deuteron polarizability to hyperfine structure as [2-3]:

$$\Delta E_{HFS}^d = \frac{4}{3}(Z\alpha)^5 \frac{\mu^3}{\pi\, m_N^2}(\Delta_1^d + \Delta_2^d), \quad (2)$$

* Corresponding author: eskin33@mail.ru

where we consider the deuteron as incoherent sum of the proton and neutron, so that $\Delta^d_{1,2}= \Delta^p_{1,2}+\Delta^n_{1,2}$. Corrections Δ_1 and Δ_2 for the proton and neutron are determined by the expressions with corresponding polarized structure functions in the form:

$$\Delta_1=\int_0^\infty \frac{dQ^2}{Q^2}\left\{\frac{9}{4}F_2^2(Q^2)-4m_N^3\int_{\nu_{th}}^\infty \frac{d\nu}{\nu}\beta_1\left(\frac{\nu^2}{Q^2}\right)G_1(\nu,Q^2)\right\}, \quad (3)$$

$$\Delta_2=-12m_N^2\int_0^\infty \frac{dQ^2}{Q^2}\int_{\nu_{th}}^\infty d\nu\,\beta_2\left(\frac{\nu^2}{Q^2}\right)G_2(\nu,Q^2). \quad (4)$$

The polarized structure functions of the proton and deuteron can be measured in inelastic scattering of polarized electrons by polarized protons and deuterons. The progress achieved in the preparation and carrying out of experiments with polarized beams of leptons, have made it possible to carry out precise measurements of polarized structure functions $G_{1,2}$ at SLAC, CERN and DESY. Spin-dependent structure functions can also be expressed in terms of cross-sections of the virtual photo-absorption on nucleons.

$$g_1(\nu,Q^2)=\frac{m_N K}{8\pi^2\alpha\left(1+\frac{Q^2}{\nu^2}\right)}\left[\sigma_{\frac{1}{2}}(\nu,Q^2)-\sigma_{\frac{3}{2}}(\nu,Q^2)+2\frac{\sqrt{Q^2}}{\nu}\sigma_{TL}(\nu,Q^2)\right], \quad (5)$$

$$g_2(\nu,Q^2)=\frac{m_N K}{8\pi^2\alpha\left(1+\frac{Q^2}{\nu^2}\right)}\left[\sigma_{\frac{3}{2}}(\nu,Q^2)-\sigma_{\frac{1}{2}}(\nu,Q^2)+2\frac{\nu}{\sqrt{Q^2}}\sigma_{TL}(\nu,Q^2)\right], \quad (6)$$

where $g_1(\nu,Q^2)=m_N^2\nu\, G_1(\nu,Q^2)$ and $g_2(\nu,Q^2)=\nu^2 m_N\, G_2(\nu,Q^2)$. To calculate the latter in this work we use interactive program MAID, which is based on a unitary isobaric model [6-7]. As a result of numerical integration (2) in the resonant and non-resonant regions, we obtain the following value of hadronic deuteron polarizability contribution: $\Delta E^d_{HFS}= 0.120\,meV$. Theoretical uncertainty of this result is determined by the errors in the measurement of polarized structure functions. The GDH sum rule for the deuteron is not satisfied with good accuracy, it also leads to significant errors. We estimate approximately total theoretical error in 50 %.

The support from Russian Foundation for Basic Research and Ministry of Education and Science of Russian Federation (grant 1394) is acknowledged (16-02-00554).

References

1. K. Abe et al., Phys. Rev. D **58**, 112003 (1998)
2. R.N. Faustov, A.P. Martynenko, Phys. Atom. Nucl. **65**, 265 (2002)
3. R.N. Faustov, A.P. Martynenko, Eur. Phys. J. C **24**, 281 (2002)
4. A.V. Eskin, R.N. Faustov, A.P. Martynenko, F.A. Martynenko, Mod. Phys. Lett. A **31**, 1650104 (2016)
5. C.E. Carlson, M. Gorchtein, M. Vanderhaeghen, Phys. Rev. A **89**, 022504 (2014)
6. D. Drechsel, O. Hanstein, S.S. Kamalov, L. Tiator, Nucl. Phys. A **645**, 145 (1999)
7. D. Drechsel, S.S. Kamalov, L. Tiator, Eur. Phys. J. A **34**, 69 (2007)

Multi-channel high-resolution terahertz spectrometer for analytical studies

V. Vaks[1,2], E. Domracheva[1,2], S. Pripolzin[1], M. Chernyaeva[1,2,*], and A. Yablokov[1,2]

[1]Institute for Physics of Microstructures RAS, 603087 Nizhny Novgorod, Russia
[2]Lobachevsky State University of Nizhni Novgorod, 603950 Nizhny Novgorod, Russia

Abstract. A method of multi-channel THz spectroscopy and a sample device that embodies the method were developed and described by the authors. The device has two independent THz radiation sources and a single receiving module. The novel method allows detecting preliminary and short-living chemical compounds and therefore to studychemical reaction dynamics. This has been shown by a series of test experiments. The method can be applied to researches in Physics, Chemistry, Astronomy, Medicine, and Biology.

The most important and fundamental studies in Chemistry, Biology, and Medicine are based on watching the ways chemical reactions follow. Studying of intermediate and short-living chemical compounds can be conducted with the method of multi-channel terahertz (THz) spectroscopy. The THz spectrometer is based on the method of transient spectroscopy that utilizes the free polarization decay within a gaseous mixture. It can be described as follows. The gas sample polarizes, under the influence of the THz radiation at the gas resonant frequency. Once the radiation generator frequency is changed to a non-resonant one, the polarized gas emits the very same radiation as the one the polarization has been induced by. This is performed via generating of a specific probing radiation. Its frequency or phase is being switched periodically. If the radiation frequency is resonant, it induces and destroys a macroscopic polarization within the gas. The spectrometer that works this way has the sensitivity close to the theoretical limit [1, 2]. Its resolution is bound by Doppler line broadening. The sensitivity remains high at a significant decrease of the sample gas pressure (up to hundredths or thousandths of a Torr); meanwhile low pressure provides high spectral resolution, as the collision broadening diminishes. The phase-switching working mode allows conducting radiation absorption measurements at a steady frequency. That is needed to study how the selected substance concentration is changing in the measuring cavity [1].

To analyse the concentration dynamics of two substances simultaneously, a new spectrometer has been proposed. The device has two independent THz radiation sources and a single receiving module [3]. The two-channel spectrometer schematic is shown at Fig. 1. There are two independent radiation sources that emit at 115–178 GHz and 330–390 GHz frequency bands. The sources are based on backward wave oscillators (BWO). The moments of their radiation phase switching are separated in time, so that the resulting gas

[*] Corresponding author: stsav012@gmail.com

polarization decays before the phase of another radiation source is switched. This results in receiving signals that are separated in time, too. As the phase switching happens quite frequently, the absorptions are measured almost simultaneously.

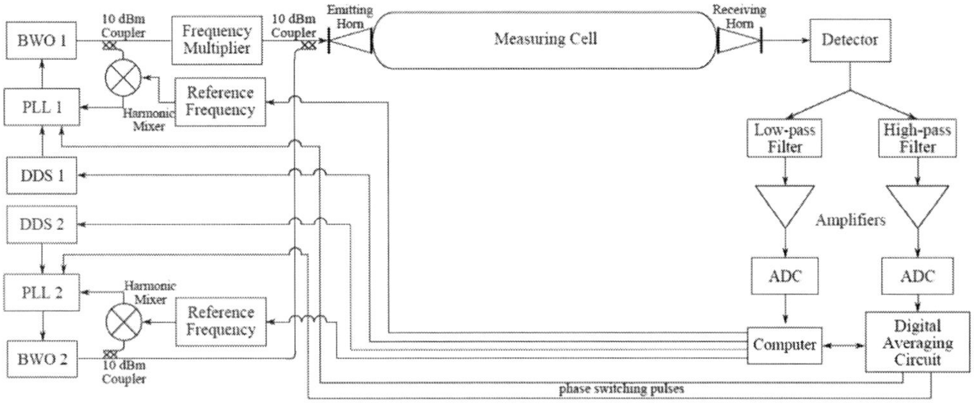

Fig. 1. The two-channel spectrometer schematic; here BWOs stand for backward wave oscillators, DDSs are direct digital synthesizers, PLLs are phase-locked loops.

We have performed test measurements using a mixture of ethanol and hydrogen peroxide. We observed a series of chemical reactions $C_2H_5OH + H_2O_2 \rightarrow CH_3CHO + 2H_2O$, $CH_3CHO + H_2O_2 \rightarrow CH_3COOH + H_2O$, and $2CH_3CHO + O_2 \rightarrow 2CH_3COOH$. They depended on a ratio of the initial compounds amounts. Besides, we have studied a mixture of formaldehyde and methanol that may be the products of some explosives slow decomposition. The measured absorption dynamics of the substances is shown at Fig. 2.

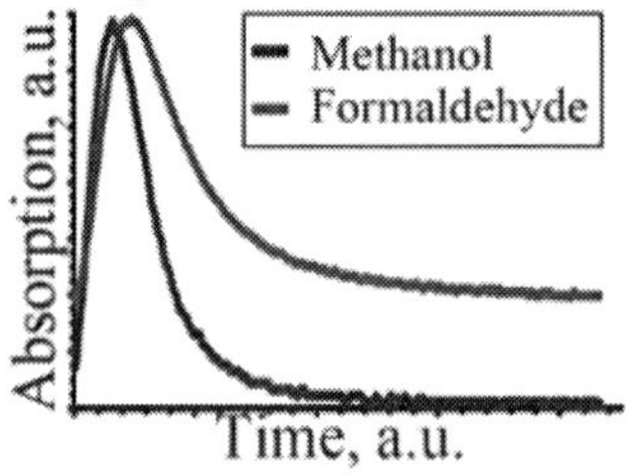

Fig. 2. Simultaneously measured absorption dynamics of methanol (bottom line) and formaldehyde (top line)

The experiments have proved that the method helps with real-time measurements of reagents concentrations. This can be applied to researches in Physics, Chemistry, Astronomy, Medicine, and Biology. The further development of the method will consist of using an IR radiation source together with THz ones. This should improve the reliability trustworthiness of the analysis and add non-polar molecules to the list of detectable compounds.

The article is the result of studies partly financed financially supported by the RFBR Grant no. 15-42-02330 and by RSF (project no. 15-12-10035).

References

1. V.L. Vaks, A.B. Brailovsky, V.V. Khodos, Infrared & Millimeter Waves **20**, 883 (1999)
2. V.L. Vaks, Journal of Infrared, Millimeter and Terahertz Waves **33**, 43 (2012)
3. V.L. Vaks, V.Y. Balakirev, M.B. Chernyaeva, E.G. Domracheva, S.I. Pripolzin, V.A. Anfertev, L.S. Revin, A.A. Yablokov, *Proceedings of An International Joint Conference of The 9th Global Symposium on Millimeter-Waves (GSMM 2016) and The 7th ESA Workshop on Millimetre-Wave Technology and Applications* (2016)

Infrared reflection spectroscopy and optical constants of $LiNbO_3$ films on crystal substrates

N.N. Novikova[1], V.A. Yakovlev[1,*], and P.G. Medaglia[2]

[1]Institute for Spectroscopy RAS, 108840 Moscow, Troitsk, Russia
[2]Dipartimento di Ingegneria Industriale, Università di Roma Tor Vergata, Rome, Italy

Abstract. We have measured infrared reflectivity spectra of thin lithium niobate films of nanometer thickness, grown by a pulsed laser deposition technique using KrF-excimer laser (λ=248 nm) on the single crystalline substrates (sapphire, MgO, NdGaO3 and SrTiO3). Using the dispersion analysis technique, we have calculated thicknesses and optical constants of the films. The phonon parameters of the substrates and films are obtained.

Electro-optical and optical photorefractive crystal of lithium niobate ($LiNbO_3$) is one of the most widely currently used ferroelectric materials in piezotechnique, quantum opto-and acoustoelectronics. Currently, active development of optical devices is based on the films of lithium niobate. The study of optical properties and obtaining the main parameters of films of lithium niobate deposited on different types of substrates, may be very promising for practical applications, pointing out ways to improve the efficiency of management of physical characteristics of this unique material. Epitaxial films of lithium niobate were grown by pulsed laser deposition on single crystal substrates of sapphire, magnesium oxide, $SrTiO_3$ and $NdGaO_3$, using a KrF (λ=248 nm) excimer laser [1, 2].

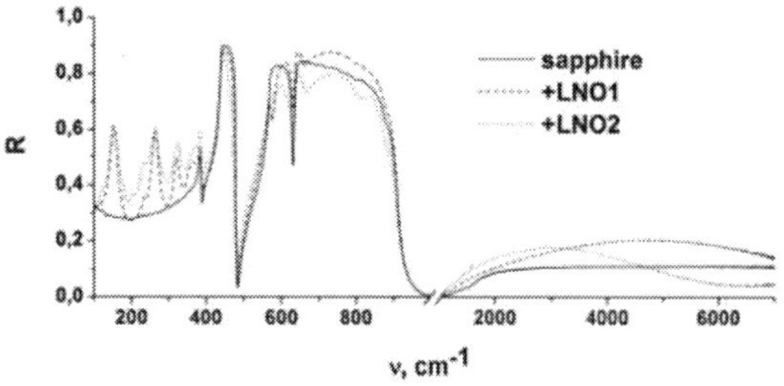

Fig. 1. Reflectivity of sapphire substrate and $LiNbO_3$ films 1 and 2.

The laser beam was focused (3 mm²) on the surface of the target ($LiNbO_3$ single crystal), with a pulse energy of 170 to 220 mJ. Typical growth parameters – pressure of O_2 is from 0.5 to 1 mbar and the substrate temperature is from 600°C to 700°C. The distance target-substrate was about 4.5 cm. After synthesis, the samples were rapidly cooled in argon atmosphere to room temperature.

Reflectance of the infrared (IR) radiation is widely used to study the vibrational spectra of bulk materials and coatings on them. This non-destructive technique allows to study the phonon spectra and to obtain optical constants and thicknesses of films. Reflection spectra in the IR region (30-7500 cm^{-1}) with a resolution of 2 cm^{-1} were registered on a Fourier-transform spectrometer Bruker IFS66v with the reflectance unit at the angle of incidence of 16 degrees. Grid polarizers on KRS-5 and polyethylene were used in the mid and the far IR regions, respectively. Some spectra are shown in Fig. 1. The optical constant of the substrates and films were obtained by dispersion analysis of reflection spectra [3,4] using the SCOUT software (Windows).

The measured IR reflection spectrum was compared with the spectrum calculated using the model of damped harmonic oscillators for the dielectric constant of the film and

* Corresponding author: yakovlev@isan.troitsk.ru

substrate. $\varepsilon = \varepsilon_\infty + \Sigma f_i^2/(\nu_{TOi}^2 - \nu^2 - i\gamma_i\nu)$. The film thickness and the parameters of the oscillators (the frequency - ν_{TOi}, the oscillator strength f_i and the damping γ_i) were varied to minimize the differences of the spectra. At first, this procedure was applied to spectra of uncoated substrates. The parameters of the substrates were obtained. They are close to the literature data (e.g. [3]). Then the film spectra were fitted taking into account the reflection from the two interfaces of the film. The film parameters were obtained. Table 1 presents the parameters of the optical phonons of the films and of the $NdGaO_3$ substrate, obtained from the reflection spectra. For comparison, the parameters of a single crystal $LiNbO_3$ from the Ref. [4] are given too.

Table 1. Parameters of the optical phonons of the films and of the $NdGaO_3$ substrate.

Film No	LNO1	LNO2	LNO3	LNO4	LNO5	LNO6	Substrate $NdGaO_3$	Crystal $LiNbO_3$ E	Crystal $LiNbO_3$ A
d, nm	256	432	682	830	485	1080			
ε_{oo}	4	4.27	4.5	4.48	4.83	5.45	4.08	5	4.6
ν_{TO1}, cm^{-1}	152.7	154.9	152.1	151	155.9	169.1	170.8	152	248
f_1, cm^{-1}	664	559	643	409	545	498	433	713	992
γ_1, cm^{-1}	19	20	13	23	18	17	7.6	2	5
ν_{TO2}, cm^{-1}	234.1	235.4	234.4	242	235.6	239	243.3	236	274
f_2, cm^{-1}	217	247	176	66	346	62	169.4	211	274
γ_2, cm^{-1}	12	11	10	7.7	11	8	7.6	3	4
ν_{TO3}, cm^{-1}	262.2	261.5	261.6	266.4	258.4	266.4	272.9	265	307
f_3, cm^{-1}	563	532	611	284	717	511	553.9	621	123
γ_3, cm^{-1}	14	15	12	15	13	9	10	3	8
ν_{TO4}, cm^{-1}	320.5	320.2	320.9	325.9	313.9	274.9	292.1	322	628
f_4, cm^{-1}	387	393	393	148	552	691	503.2	478	1000
γ_4, cm^{-1}	14	15	8	8	40	30	11.6	3.5	22
ν_{TO5}, cm^{-1}	361.8	360.2	362	404.4	321.6	318.5	338.7	363	692
f_5, cm^{-1}	454	474	727	655	236	250	553.6	551	250
γ_5, cm^{-1}	41	41	42	32	8	12	18	12	34
ν_{TO6}, cm^{-1}	426.9	418.3	432	443.5	362.9	343.9	442.3	431	
f_6, cm^{-1}	511	505	237	558	438	648	91.6	183	
γ_6, cm^{-1}	40	42	23	41	23	44	6.2	5	
ν_{TO7}, cm^{-1}	565.1	565.8	556	558.7	433.3	584	521.8	586	
f_7, cm^{-1}	891	923	289	912	197	724	124.5	1065	
γ_7, cm^{-1}	24	56	30	57	21	59	22.6	21	
ν_{TO8}, cm^{-1}	585.4	576.2	561.8	600.7	591.7	621.9	594.3	670	
f_8, cm^{-1}	501	291	859	307	1106	320	232.3	300	
γ_8, cm^{-1}	7	32	55	30	60	23	22.9	31	
ν_{TO9}, cm^{-1}	668.7	670.1	670.1	674.2	655.5	668.6			
f_9, cm^{-1}	176	325	325	340	347	316			
γ_9, cm^{-1}	22	69	69	71	69	88			
Substrate	Al_2O_3	Al_2O_3	MgO	MgO	$SrTiO_3$	$NdGaO_3$			

References

1. G. Balestrino et al., Appl. Phys. Lett. **78**, 1204 (2001)
2. M.A. Tagliente et al., J. of Crystal Growth **216**, 335 (2000)
3. A.S. Barker Jr., Phys. Rev. **132**, 1474 (1963)
4. A.S. Barker Jr., R. Lowdon, Phys.Rev. **158**, 433 (1967)

IR absorption spectra in the region of stretching and deformation vibrations of hydrogen bond in pure and alloyed single crystals of $LiNbO_3$

N.N. Novikova[1], V.A. Yakovlev[1], and N.V. Sidorov[2,*]

[1]Institute for Spectroscopy RAS, 108840 Moscow, Troitsk, Russia
[2]I.V.Tananaev Institute of Chemistry and Technology of Rare Elements and Mineral Raw Materials, Kola Science Center RAS, 184209 Apatity, Russia

Abstract. We have measured infrared absorption spectra of the stretching and bending vibrations of the hydrogen bond of pure and alloy additives Me (Me = Mg, Zn) single crystals of lithium niobate ($LiNbO_3$) grown using various technologies. Preliminary studies of the spectra showed that the frequency of stretching vibrations in the IR spectra of single-crystal samples obtained from the homogeneously doped charge differ significantly (more than 50 cm^{-1}). However, no difference in the IR spectra of single crystals obtained using other doping techniques, we have found. It is assumed that the observed differences in the absorption spectra of single crystals due to the different character of the formation of hydrogen bonds in the preparation of precursors Nb_2O_5: Me.

The infrared absorption spectra were studied in the polarized radiation in the region of stretching and deformation vibrations of hydrogen bonds for more than 60 nominally pure and alloyed in a wide range of concentrations of alloying elements Me (Me=Mg, Zn) single crystals of lithium niobate ($LiNbO_3$), grown by various techniques. Four techniques of doping were used to prepare these crystals:

1. Homogeneous doping is the synthesis of granulated mixture with a high bulk density is produced using a homogeneous solid Nb_2O_5 :Me precursors, obtained by homogeneous doping of Me, recovered at the stage of extraction separation Nb_2O_5. This method of doping is based on sorption of respective cations of alloying of high-purity niobium hydroxide. At the same homogeneously doped solid precursor Nb_2O_5:Me, in turn, was obtained in two ways: by separation from reextract, obtained by extrakt processing or fluoride-sulfuric (method I) or fluoride-hydrochloric acid solutions (method II) from processing rare metal raw materials and wastes of lithium niobate.
2. Direct addition - direct addition of the appropriate alloying of the Me oxide into the granular mixture with a high bulk density before fusion of the crucible.
3. Addition before fusing the crucible into the granular mixture with a high bulk density of the synthesized $LiNbO_3$-based special alloys, enriched in alloying elements Me.

* Corresponding author: sidorov@chemy.kolasc.net.ru

4. The use of granulated furnace charge with high bulk density, synthesized from a homogenous mixture of niobium pentoxide, the oxide of alloying element Me and lithium carbonate.

All the single crystals both nominally pure and doped, were grown in air atmosphere by Czochralski method using the same methodology on the "Crystal-2". Crystalline samples for studies had the shape of rectangular parallelepipeds with dimensions of ~ 7x6x5 mm^3 with edges coinciding in direction with crystal axes X, Y, Z, where the Z axis is the polar axis of the crystal. The faces of the parallelepipeds were carefully polished. Preliminary studies of the IR absorption spectra of single crystals obtained by different techniques showed, that in the region of stretching and deformation vibrations of hydrogen bonds there are significant differences in the frequencies of the oscillations. So the frequency of stretching vibrations in the IR spectra of single crystal samples obtained from a homogeneous alloy mixture, in turn obtained using Nb_2O_5: Me precursors by method I and method II, are significantly (more than 50 cm^{-1}) different. However, no differences in the IR spectra of single crystals obtained using other methods of alloying, we did not found.

The detected differences in the IR absorption spectra of single crystal samples is, in our opinion, due to the different nature of the hydrogen bonds formed in the preparation of precursors for Nb_2O_5:Me in fluoride-sulfuric acid (method I) or fluoride-hydrochloric acid solutions (method II). Study of the nature of the hydrogen bonds formed in these two types of solutions is a matter of further special study.

Reference

1. M.N. Palatnikov, I.V. Biriukjva, O.V. Makarova, N.V. Sidorov, N.A. Tepliakova, S.M. Masloboeva, V.V. Efremov, Journal Perspektivnye materialy **1**, 5 (2016) (in Russian)

Infrared reflection spectra of the films of topological insulator $Pb_{1-x}Sn_xSe$ on the substrates ZnTe/GaAs

N.N. Novikova[1], V.A. Yakovlev[1,*], I.V. Kucherenko[2], G. Karczewski[3], and S. Chusnutdinow[3]

[1]Institute for Spectroscopy RAS, 108840 Moscow, Troitsk, Russia
[2]Lebedev Physical Institute RAS, 119991 Moscow, Russia
[3]Institute of Physics, Polish Academy of Sciences, PL-02668 Warsaw, Poland

Abstract. Infrared reflectivity spectra of the 700 nm thick topological insulator $Pb_{1-x}Sn_xSe$ films grown by the molecular beam epitaxy technique on ZnTe/GaAs substrates were studied. Using dispersion analysis of reflectivity spectra plasmon and phonon parameters for the samples under study were obtained.

Narrow-gap alloys type IV – VI , such as PbSnSe, PbSnTe has generated great interest from fundamental and applied points of view in the 70s-80s years of the last century. Changing the concentration of Sn and ambient temperature, it is possible to vary the size of the forbidden gap in the L-point of Brillouin zone, reaching a gapless state. The dependence of E_g on the temperature T and the composition (x) is given as:

$$E_g\,(\text{eV}) = 0.13 + 4.5x10^{-4}\,T - 0.89x \qquad (1)$$

According to this formula, the values of E_g can be both positive and negative. This means that in the first case, the wave functions describing the conduction band and the valence zone at the L point have the symmetry of L- and L+. In the second case, the inversion zones and their symmetries correspond to L+ (the conduction band) and L- (valence area). Recently [1], in the study of the reflection spectra of ARPES investigation, it was found that the surface of PbSnSe compounds with inverse location areas in the volume has a status Dyakowski range. Such compounds have now been attributed to topological insulators.

We studied the infrared reflection spectra of thin films of different composition of a topological insulator $Pb_{1-x}Sn_xSe$ with a thickness of 700 nm, grown by molecular beam epitaxy on the substrates ZnTe/GaAs. Studies were conducted on four different samples with $x = 0.20$, 0.22, 0.34 and 0.40. The samples were grown at the Institute of Physics PAN. Spectroscopy reflectance infrared (IR) radiation allows to study the phonon spectra and to obtain optical constants and thicknesses of the films. Reflection spectra obtained with the infrared Fourier transform spectrometer Bruker IFS 66v/s at the angle of incidence close to the normal in the spectral range from 50 to 7500 cm^{-1} and at a spectral resolution of

* Corresponding author: yakovlev@isan.troitsk.ru

DOI: 10.1051/epjconf/201713203059

4 cm^{-1}. The optical constants of the substrates and the films, their thicknesses and complex permittivities ε were obtained by the dispersion analysis of reflection spectra [2] using the software SCOUT [3] for four films of different composition $Pb_{1-x}Sn_xSe$. The measured IR reflection spectra (Fig. 1) were compared with the spectrum calculated using the model of damped harmonic oscillators for the dielectric functions (2) of films and substrates, taking into account plasmon as well

$$\varepsilon = \varepsilon_\infty + \Sigma f_j^2/(\nu_{TOj}^2 - \nu^2 - i\gamma_j \nu) - f_p^2/(\nu^2 + i\gamma_p \nu) \qquad (2)$$

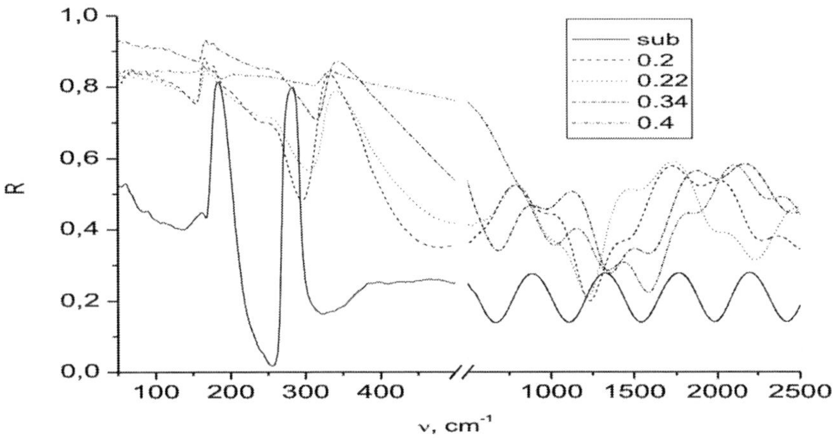

Fig. 1. Spectra of the substrate and of four $Pb_{1-x}Sn_xSe$ films.

The film thicknesses and the parameters of the oscillators (the transverse phonon frequency - ν_{TOj}, the oscillator strength f_j and the damping γ_j) and plasmons (f_p and attenuation γ_p) were varied to minimize the differences of the spectra. At first this procedure was applied to the spectrum of the clean substrate (ZnTe film on GaAs plate). The parameters of the two materials and the thickness of the ZnTe film were obtained. Then the spectra of the films were fitted taking into account the reflection from all the interfaces. The film dielectric functons are shown at Figs. 2 (Re ε), 3 (Im ε) and 4 (Im ε^{-1}).

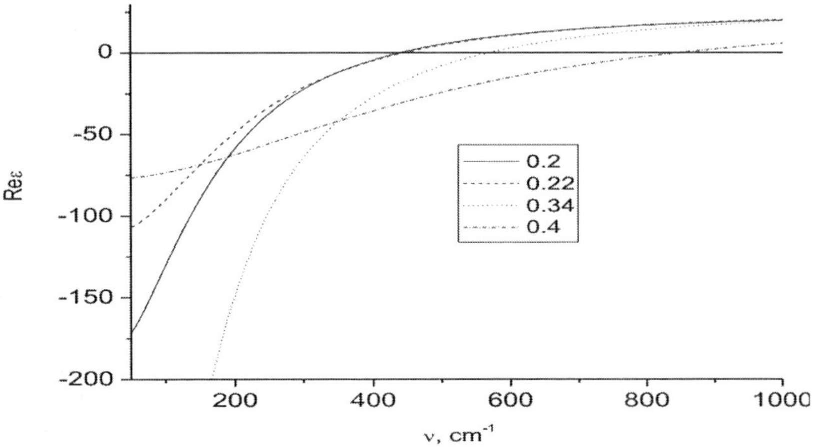

Fig. 2. Re ε of four $Pb_{1-x}Sn_xSe$ films.

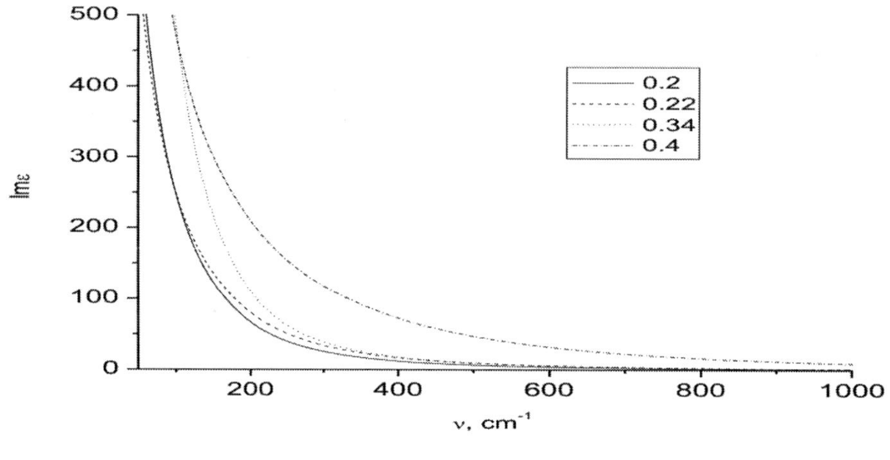

Fig. 3. Im ε of four $Pb_{1-x}Sn_xSe$ films.

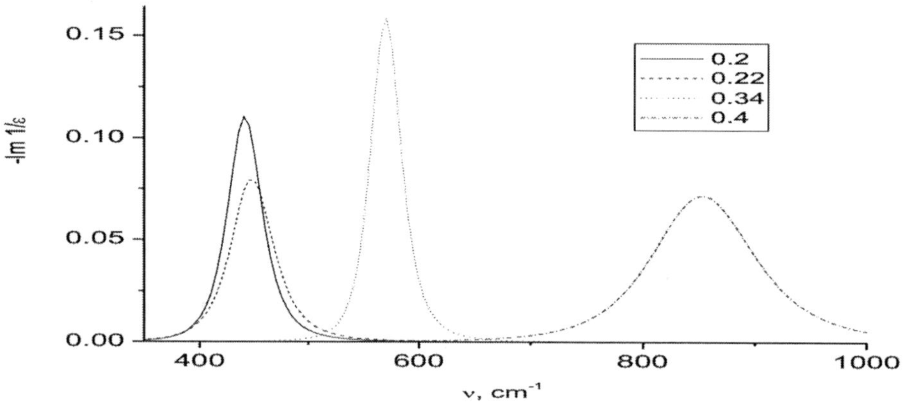

Fig. 4. Loss functions of four $Pb_{1-x}Sn_xSe$ films.

Negative values of Re ε (Fig. 2) and high Im ε (Fig. 3) show that the main contribution into the film dielectric functons is due to the film plasmon (formula 2). The maxima at the loss functions (Fig. 4) spectra show the film plasmon frequencies. $Pb_{1-x}Sn_xSe$ phonons are lower (around 40 cm^{-1}).

This work was partially supported by the Foundation of Polish Sciences by the Master program № Mistrz5/2013.

References

1. C.M. Polley, P. Dziawa, A. Reszka, A. Szczerbakow, R. Minikayev, J.Z. Domagala, S. Safaei, P. Kacman, R. Buczko, J. Adell, M.H. Berntsen, B.M. Wojek, O. Tjernberg, B.J. Kowalski, T. Story, T. Balasubramanian, Phys. Rev. B **89**, 075317 (2014)
2. A.S. Barker Jr., Phys. Rev. **132**, 1474 (1963)
3. W. Theiss, *The SCOUT through CAOS. Manual of the Windows application SCOUT* (1994)

AUTHOR INDEX

AUTHOR INDEX

AUTHOR INDEX